THE REAL ORIGIN OF THE UNIVERSE

By Francisco Sanchez, JR

© 2019, Francisco Sanchez, JR

All Rights Reserved

No part of this book may be copied, sold or distributed, in either printed or electronic format, without the written permission from Francisco Sanchez, Jr. For more information, contact Forastero21amado@gmail.com.

Kindle Edition

About The Real Origin of the Universe

The Real Origin of the Universe is about the true and the only origin of the universe. The tall title really says it all.

Those conscious beings that do not know reality or do not know the truth thus they will not be able to live in reality or in the truth nor will they live as reality or as the truth they them very selves and they live as if they do not live.

And in truth, they will die because of the lack of reality or because of the lack of the truth herself, that if recognizing reality or the truth herself will give them true or royal life and in true or royal abundance of peace, knowledge, joy and gladness...

Tags: Origin, universe, cosmos, existence, matter, space, black holes, space-time, atom, element, table of elements, periodic table of the elements, electron rings, magnet, bar magnet, magnetic field, protons, neutrons

About this very successful Author Beloved

Francisco Sanchez, Jr. is one of those, and perhaps the only one in these very difficult times, who with complete sincerity and humbleness, gladness and joy really searches and researches the truth of the truth, the truth of the universe, the truth of creation and the truth of what is the rebirth of man or of mankind...

Now then, every conscious man should have a copy of this book even though he needs not be a physicist or any other scientist or even a searcher or researcher but a glad and joyful contender who contends for the truth...

Table of Contents

Prolog..........7
Dedication..........8
Introduction..........9
Chapter One..........41
Why is the universe round? 41

Chapter Two..........60
Dimensions of time..........60
The Stars and the Black Holes..........68

Chapter Three..........73
The Periodic Table of the Elements..........73
The Elements and Their Weight..........85

Chapter Four..........86
The Divisible Atom..........86
The Small which makes the Great..........95
Creation and the Number..........98

Chapter Five..........104

Summary..........104

Something More About Existence..........

The More Understanding of Existence one has..........114

The Plus Zero Negative Factor..........125

Conclusion..........127

Something about the Author..........136

Who am I really? 138

Prolog

Truly glad and truly joyful is that man that has true knowledge of how existence or of how the universe truly functions because that man also knows how that man himself truly functions.

And he also will do as existence her very self to continue on existing or living on but existing or living on as if forever new and in complete abundance...

Thus, Know the function of the real universe or the function of existence and you will function much better or with a greater identity, because to come to know was because of doing and thus entering into understanding, which really is to enter into the right side of existence as the right side of existence and thus one becoming as the very beginning because of one renewing or because of one refreshing existence because of one knowing or one understanding...

Dedication

The Real Origin of the Universe is dedicated with all my heart and with my entire conscious mind to the good reader and to those who gladly and joyfully fight or gladly and joyfully struggle to understand or to find the truth and enter in the truth and thus become the truth herself.

Introduction

Explaining the origin of the universe, the origin of creation, the origin of existence or of reality or even the origin of time, and the physical inner and the physical outer workings, is not an easy task, but it can really be done!

In fact, it has truthfully and completely being already done and it did not even take eight to nine hundred pages or even more!

And explaining and thus proving the origin of the universe, the origin of creation, the origin of existence or of reality or even explaining the origin of time, and its physical inner workings and its physical outer workings, does not really require a couple or more university degrees or a fancy title or two or even three as it does not really require those multiple degrees or those fancy titles to completely understand and thus completely enjoy the origin of the universe, the origin of creation, the origin of existence or the origin of reality or even of time, and its physical inner workings and its physical outer workings!

In short, one does not have to be a genius or be a wizard to fully understand and to completely enjoy the true or the naked origin of the universe or of creation!

Naked because the origin or the truth is but naked! The naked truth bears all or bears all of existence or bears all of reality or creation. There is nothing or something really hidden!

Everything really or actually is before the naked human eyes, but the more the knowledge of reality or of actual existence or even of creation, the more understanding and thus the more of reality or the more of existence or of creation will be seen!

That is to say, the more true knowledge of existence or of actual creation or the truer knowledge of creation or of the universe, the more actual existence or the more creation or the universe will be seen and understood.

Most scientists, most philosophers or even most educators believe because a scientific theory or an educated guess sounds good, that makes the good sounding theory or the educated guess true even though the good sounding theory or the educated guess really or actually cannot be proven.

But the naked truth, which is but complete or but full or perfect knowledge, is simply but very obvious.

And that which is but very obvious simply is and can really be proven. There is no need to add color or no need to add sound to the naked truth or to existence or to creation herself, doing so is hiding or coloring or simply is tinting what really or actually is!

In other words, being naked is so obvious that it speaks for itself and it makes no sense to say that it is naked when it already is!

Once again, one simply does not really have to be a genius or really be a wizard to fully understand and to completely enjoy the true or the naked origin of actual existence or of actual creation!

Anyone who can simply add one plus one and get a real and very satisfying answer or anyone who simply knows his name and the simple meaning or the simple definition of that simple name can simply understand and can completely enjoy the physical and the actual origin of the universe and of existence or of creation!

Sadly, most people do not really or do not actually know the full or the complete meaning of their names and thus do not really enjoy the meaning and do not enjoy its grandiose power!

The requirements to completely understand and to completely enjoy the true physical or the naked origin of the universe, of creation, of existence or of actual reality or even of time, and its true physical inner and its true physical outer workings are very simple and are very few: to be interested enough to simply form, to simply create or to even simply ask a very simple question, a correct and answerable question!

That is to simply say, to be personally interested enough to ask a very simple question in a certain and in a simple way that the very simple question can be can be completely answerable or can be completed and understood and thus be enjoyed!

To position or ask that very simple question in a certain way that the very simple question can be completely answered and completely be understood be enjoyed.

In other words, simply or just built or just create a simple solid foundation of actual reality with a very simple question!

That is to say, depart from physical reality and physical reality will be seen!

The human brain, for example, will simply complete or will simply fill that which the naked human eye actually or really cannot see, but first "that" must be looked at and seen!

Draw two simple lines together or intersecting one way or another and the human brain will add a third line, trying, therefore, to make something complete or something meaningful of the first two simple lines!

Draw a half circle and the human brain will try to make something of that half circle or add to that half and simple circle, trying to make that simple circle full or compete and understandable or useful and thus enjoyable.

Too often, therefore, a completed picture or a completed diagram or a very simple drawing can only be seen or can only be fully understood and fully enjoyed when there is a bit of physical motion of the naked human eye or a very simple physical repositioning of the interested observer or the interested learner.

And even though the physical motion is not from the simple picture or from the very simple diagram or from the very simple drawing, that simple physical motion or point of view or the very simple physical repositioning of the interested observer or the interested learner truly helps to complete, to simply see or to fully and simply understand and thus fully enjoy the complete picture or the complete diagram or the very simple or complete drawing!

It is as if the physical motion or the point of view of the naked human eye or the simple personal interest or the very simple physical repositioning of the interested observer or the interested learner fully completes or simply complements the simple picture or completes or complements the very simple diagram or the very simple drawing even though the simple diagram or the simple picture or the very simple drawing truly was already complete or was already perfect.

In other words, a bit of simple physical motion or change or improvement of point of view or the very simple physical repositioning of the naked human eye helps to simply and to fully understand and thus enjoy fullness or really enjoy actual reality!

That is to simply say, the very simple personal point of view or personal interest or the very simple physical repositioning of the very simple observer helps to simply complete or to simply make perfect existence and thus enjoy existence in and through simple enjoyment!

Explaining the true origin of the universe or of creation or of existence or of reality or even the true origin of time, and its true physical inner and its true physical outer workings, requires both a simple theory or some personal interest or point of view and some solid or some physical scientific facts or solid bits and pieces.

And it also really or actually requires describing something solid or something physical and something simple that the human being cannot really or actually see in and out in its entirety (with some of the simple and solid pieces or some solid scientific facts or solid bits and pieces that the conscious or that the simple thinking human being already has or simply thinks or simply believes that he may have but sadly has not).

In other words, it takes some physical reality or some solid bits or some solid pieces of physical existence or actual reality or physical creation to further explain physical reality or physical existence or physical creation and thus explain the simple physical origin of the entire and the physical universe and physical or actual creation!

It takes solid or physical facts or solid bits and pieces with some personal interest or good or better point of view to simply and to truthfully explain physical existence or actual reality or creation and its true inner and its true outer workings!

The simple personal interest or point of view is the simple physical and the actual motion that will physically put together the solid or the physical facts or the solid bits and pieces and, therefore, complete the physical universe or complete physical existence or complete or finish actual reality or complete physical acknowledge creation!

The human being, for example, cannot simply see his own physical face or see his complete or true physical face in front of him unless there is a solid or a simple physical reflection or a real or actual physical sensation of some kind!

But when he finally gets to see a solid or a physical refection or feels his solid face or complete or true physical face, the human being can physically see or can physically sense or can even physically feel the exact physical opposite of what he is really or actually feeling, really or actually sensing or even really or actually looking at!

And it is really or actually easier to simply theorize or it is easier to simply fantasize than it is to truthfully or to completely believe or to truly or to completely see or to fully and to simply understand physical reality as physical reality or as physical existence really or actually is and how physical reality or how physical existence really or actually works.

The true or the physical reality, or simply the naked truth, and the sad irony of the physical and the whole matter is that the conscious human being will never be able to see his true or real and physical face or complete face as his true or real and physical face or complete or actual face really is in physical reality or in physical existence no matter how good and simple the solid or the physical refection or how great the real and the simple physical sensation!

It is not because of constant physical change or constant physical motion or that the conscious human being simply and physically grows old, gray and blind that he cannot simply and thus really see his true face or complete or actual face, but because the conscious human being, as of yet, cannot really see or cannot fully understand physical reality or cannot understand physical existence as physical reality or as physical existence really is!

And not only that, but enjoying physical reality or really enjoying physical existence is also a major part of physical reality or physical existence, no matter how physically limited the actual or the present reality or the actual or the present and physical existence!

A complete, a perfect or a full face is a simple face with a simple or a with great or joyful smile, but once that simple and great or joyful smile is gone, so is the complete, the perfect or the full face!

Therefore, to have a complete, a perfect or a full face, the conscious living being must also and simply have a complete, a perfect or a full mind!

A complete, a perfect or a full mind is a mind with complete, with perfect or with full or with true knowledge of reality or actual existence or even creation!

True, simple or naked knowledge of reality or actual existence or creation is that knowledge that not only brings forth more true or more naked knowledge, but true or naked knowledge also expands the human mind through complete or through full joy or through total or through complete or perfect enjoyment!

In other words, naked or complete knowledge of actual existence or of physical reality or of actual creation really opens or really gives more visions to the naked human eyes so that the conscious human being or the conscious living being can really or can actually see the naked, true or complete physical reality or actual creation that already was or that really or actually is in front of the conscious living being.

The simple function, the simple purpose or the simple job of the natural human mind or the simple function or the simple job of the natural human receiver, or simply call it the simple animal instinct or the simple animal mind, is to see in simple solid parts, in solid bits or in solid pieces, and not see whole or complete!

The simple animal mind simply cannot see in full or cannot see in fullness. The simple animal mind or the simple animal instinct cannot really see or cannot really understand in fullness.

The human being, for example, cannot see the whole or cannot see the complete picture or cannot see the whole or cannot see the complete diagram or cannot see the simple drawing unless the human being first sees simple physical parts or first sees physical or solid pieces of that whole or that complete picture, or whole or complete diagram or simple drawing, or the whole or the complete picture or the diagram or that simple drawing is in simple and physical motion or the whole or the complete picture or the diagram or the very simple drawing smells and then the human being simply reacts or simply takes notice.

It is also an animal instinct to see in simple physical or in a solid part or in simple solid pieces and to physically react to the simple physical motion or react to the simple physical sound of the simple physical or actual motion or react to the new point of view.

The simple animal or the simple natural mind or the simple natural instinct, for example, cannot really or cannot actually see a physical whole or really cannot see a complete tree.

That is to say, the simple animal mind or the simple animal instinct does not allow the animal to see a whole or a complete tree.

An animal mind can only see solid parts of that physical tree; and even though that animal mind can see different solid parts of a physical tree, that simple animal mind cannot really see the whole or cannot see the physical and the actual tree!

In the same manner, animals or insects, no matter how smart or even well trained, can only see light and shades.

Again, the simple animal mind or the simple animal instinct or really point of view cannot really see the physical sun, cannot see the physical stars or even cannot see the physical moon, but can only see the physical light and the physical shade.

It is through the simple shade and the simple flickering or the simple bit of physical motion that the human being through his simple animal or natural mind or through his point of view can simply see physical reality or can really see physical existence as physical reality or as physical existence or even physical creation really is!

The solid or the physical part through some personal interest or through some physical motion or point of view really explains the solid or the physical whole.

The solid or the physical whole really cannot or actually cannot be fully or completely seen or cannot be completely understood unless there is some physical motion or there is something physically doing the simple addition, the simple division or the simple break-up!

To simply describe or to simply analyze something physical or physically, the scientist, the researcher or the writer, or even a police detective, therefore has to simply either physically take it apart or to simply put it physically together from the simple and solid bits or from the simple and solid pieces the scientist, the researcher or the writer, or even a police detective, already has or the simple solid bits or simple solid pieces he sees or understands!

From the missing solid bit or from the missing solid piece or from the missing solid pieces or from the missing solid bits the scientist, the researcher or the writer, or even a police detective, simply theorizes or simply makes an educated guess or develops a new point of view, thus making simple physical reality or making the simple solid fact or the simple naked truth very difficult to simply see or very difficult to simply or really understand and very difficult to simply prove and thus enjoy.

The simple theory, the simple educated guess or the simple personal interest or the new point of view is, therefore, the simple glue to simply glue together the simple solid bits or the simple solid pieces or the simple physical reality or the simple solid facts or the simple naked truth that the scientist, that the researcher or the writer, or even a police detective, already has, but too often the simple glue or the simple theory or the simple reasoning or the simple educated guess or point of view is more than the simple reality, and ironically makes but more puzzles or more solid pieces than those that the scientist, the researcher or the writer, or even a police detective, already has or actually or really started with!

What a sticky mess!

And so most writers and most researchers as most scientists, sadly but true, become more interested in the good sounding or the good smelling glue, therefore filling countless books with half-truths or with false knowledge having nothing to do with the simple physical reality, the simple naked fact or the simple solid pieces that the writers, the scientists or even the researchers actually started with!

What sad irony and what a waste of valuable time and what a waste of precious physical resources! Confusion is simply not knowledge, but confusion is simply the lack or the simple misunderstanding or the simple misuse of knowledge!

That is to say, true or real knowledge is physical and actual reality; and the lack of true or real or naked knowledge is not physical or is not actual reality.

In other words, complete or full or perfect knowledge is complete or is full and physical reality. And complete or full reality is but complete or full enjoyment; and full enjoyment simply brings an expansion or an abundance of conscious mind!

And that is the simple and the actual reality or the simple naked truth!

In other words, real or true knowledge, or simply the simple naked truth, really and simply brings forth a simple but very useful expansion or brings forth a simple but very useful abundance of conscious mind or brings forth an expansion or an abundance of conscious thought.

An expansion or an abundance of conscious mind or conscious thought brings forth more real, truer or more very useful naked knowledge. But that is simply not happening!

Even well trained or highly experienced scientists, researchers or professional writers are still very prone or very sensitive to their own point of view, simple personal prejudices, to their simple or complex personal problems, to their simple cultural or even to their simple religious beliefs or to their simple and lack of solid scientific or the lack of solid information or the simple lack of solid or naked knowledge or even the very simple misunderstanding or the very simple misuse of true or naked or physical knowledge!

For example, which of the two following questions has a real or has a possible or complete answer?

Which of the two following questions really or truthfully has a solid or has a physical answer?

Which of the two following questions really has a solid or has a physical foundation?

"Why is the Universe round or sphere-like?" Or "What makes the Universe round or sphere-like?"

The "Why?" or the very first question is a very childish, is a very foolish or is a meaningless and a useless question!

The "Why?" question really limits point of view or really limits way of thinking. The very childish, the very foolish or the meaningless and useless question is asked as if no real, no actual or no true physical or complete answer is really expected from the person asking the very childish, the very foolish or the useless and thus meaningless and incomplete question!

At the very same time, that very childish, that very foolish or that useless and thus meaningless question keeps the person from simply seeking an actual or a real or a physical or complete answer simply because that very childish, that very foolish or that useless and thus that meaningless question sounds very complicated, sounds very unreal or sounds simply unanswerable!

And if the very childish, the very foolish or the useless and thus meaningless and unanswerable question were actually or really asked to another person, that other person too would be dead stuck and would be dumb founded, and would not even attempt or would not even bother or simply would not be personally interested enough to even give an answer, any answer for the answer too would be very childish, very foolish, totally meaningless, and without any real understanding and thus without any real or without any true or without any meaningful enjoyment!

The "Why?" or the very first question is like an adjustable pipe wrench or like a simple tool deliberately thrown into the gear box that is the human brain!

In other words, the "Why?" question, if it is actually or really a question, is a very simple or a very silly joke that no one simply or really understands and because no one really understands the silly joke, no one can simply or really appreciate the silly joke and simply or really cannot enjoy the silly and meaningless joke!

Remarkably, however, that very childish, that very foolish or that useless and thus meaningless and unanswerable question is asked by most scientists, by most writers and by most professionals than it is simply and really asked by a very funny or a very silly or a very humorous child or simply asked by very curious and very playful children!

On the other hand, "What makes the Universe round or sphere-like?" has an answer, a very simple and a very satisfying, understandable and thus a very enjoyable and a real and a physical and complete answer!

In fact, part of the very simple and the very satisfying and very understandable and real and physical answer is already in the very simple and the answerable question because the simple question is a real or is a complete question!

The simple and the answerable and the real question is already telling us that something is, therefore, really making the Universe round or sphere-like!

It is a very simple but true question that although will lead to many more simple and many more real or true questions, the simple questions will also give one final, one simple and one truthful or one complete and one very satisfying answer, an answer that can be fully or can be completely understood and thus simply enjoyed!

However, it does not really take too much creativity, too much imagination or too much inventiveness or too much glue or even too much of a theory or educated guess or even point of view to simply see, to simply understand or to simply discover and thus simply enjoy the naked truth or the simple physical or solid reality or physical creation that is already in front of mankind or already was in front of human kind since the very beginning of conscious thought or point of view!

The very simple and the naked truth or the very simple physical or solid reality, as it was already and very simply stated above, and its simple inner workings and its simple outer workings, is simply in front of or before mankind.

In fact, mankind walks right through physical or complete reality every day; but because mankind already is a physical part of what mankind is very hard trying to explain, to see or to complete or to simply understand, mankind simply cannot see the whole or cannot see the complete picture, especially if mankind is the one doing the simple physical motion or doing the constant but simple change or continuation!

And because mankind usually and really questions or usually and really asks for the wrong reason or reasons, mankind simply cannot see in the simple light or even sense the simple light and cannot walk in the simple light. Mankind becomes more of what he does not really want to be with the very and the simple knowledge or the solid pieces he thinks that he may have, but ironically he really has not!

And what is that simple and the naked truth or the simple physical or the solid reality or the hidden and simple physical truth, and its simple inner workings and its simple outer workings, that already is before or in front of mankind?

The very simple reality or the very simple and complete or naked truth is that mankind does not let his very own brain do its wondrous and enjoyable job!

The human brain or human conscious thought is more than a simple copy or more than a simple duplicate of physical existence or physical reality or even creation. Mankind uses his conscious mind, a very small portion of the human brain, to simply misinform or to even poison and to cripple on purpose the human brain!

One way or another, mankind on purpose limits his point of view or really covers the simple reality or the simple naked truth or filters out the proper physical information or filters out the proper or the true knowledge of reality or actual existence or creation that will let the human brain to simply balance itself with true or with complete physical reality or with true or with complete physical existence or actual creation, which simply is in constant but simple change or continuation!

The human being purposely and simply incases or imprisons his human brain by purposely and simply incasing or imprisoning his human brain by feeding the human brain false knowledge of reality or actual existence or creation or by feeding the human brain half-truths or by feeding the human brain poison!

In other words, if the human brain simply does not have the proper physical or simply does not have the correct information or the proper or the correct recall or the proper or the correct memory of actual reality or physical reality or actual existence or creation, there is but total physical chaos. The human brain is already physically dead before the due time!

For example, when a physical part of the human body no longer reacts to physical pain or to physical sensation or the physical body part is motionless, not giving proper or correct physical information or proper or correct physical feed back to the human brain, the human brain will physically shut off that misinforming part of the human body!

To the human brain, therefore, correct or true physical information or complete or truth knowledge of actual reality or physical existence or creation is more important than is a malfunctioning or misinforming body part or even a misinforming human mind!

And that is the simple and the naked truth or the physical reality thus far!

What, then, does this entire and simple introduction have to do with the true physical origin of the entire Universe or true reality or true creation, and its true inner and its true outer workings, one might simply ask?

Ironically, the most difficult part of this simple scientific theory or this simple work, or this simple glue, has being the introduction. Physical lack of true or complete knowledge or lack of knowledge of actual reality or of creation is not the only thing that keeps mankind or the human race physically and mentally from moving progressively or successfully forward in time or in history, but also that which mankind thinks or believes that he simply has but actually has not!

The sad reality or the sad fact or the sad and naked truth is that the human race has followed the very same simple line or the very simple pattern of thinking or the very same reasoning or point of view from as far back as 6,000 to 12,000 years or more!

The so-called human or the so-called scientific knowledge or point of view simply does not really add to physical much.

The human race or mankind really or actually is still living in the dead past, with the very simple physical exception that his modern brain shuts off parts of the his human body simply because of lack or very poor information, or simply because of lack or very poor solid facts or the lack of physical information or poor point of view of actual reality, of physical existence or the simple lack of information of simple human surroundings, natural or unnatural!

And knowing the true physical or the true origin of the entire Universe or of creation or of all physical existence or even knowing the true origin of physical time, and its simple inner workings and its simple outer workings, will not really or simply will not add to physical much or to anything unless mankind simply sees or simply knows or simply understands completely or in full what actual or physical reality or actual or physical existence or creation is all about.

And once the conscious living being really or actually sees reality or creation as reality or as creation really is, reality or creation is no longer the same.

Once again, the very simple reality or the complete truth or the naked and the very simple truth, lets call it that, is in front of mankind.

In fact, mankind walks right through actual or physical reality every day. But because mankind simply and usually questions, senses or feels, or even fears, for the wrong reason or reasons, thus mankind simply cannot really see in the simple light.

Of all the living creatures, man does not only break all the physical laws of physical nature, but also man is the worst interpreter of all the physical laws of physical nature!

In fact, if man, from the very beginning of consciousness, really understood or really recognized the simple naked or the simple physical laws of physical nature, none would have being broken, and man would be so much ahead in conscious mind and in physical body!

Sadly, however, mankind simply becomes more that complicated physical law breaker or that confused physical creature or that useless or that complicated animal or that complicated or useless physical thing that mankind so much try so hard not to simply be.

Sadly, mankind is simply becoming an unnatural animal or a physical creature of solid but useless stone or useless mud or broken clay!

Mankind is becoming a cripple and useless clay animal that cannot even and simply drag itself into the simple light!

As it can be simply seen, the very simple and the actual physical reality or simply the simple naked truth is that mankind, once again, simply does not let the human brain to simply function!

That is, the human being does not let the human brain do the brain's wondrous and very simple and enjoyable job!

The human brain simply is more than a simple physical duplicate of actual existence or actual or physical reality or even creation herself!

Mankind uses his simple conscious mind, a very small portion of the human brain, to deliberately misinform or to deliberately poison the human brain!

The human brain, being a simple duplicate of physical existence or creation, has true or has complete or has full knowledge of all physical existence or all physical reality!

One way or another, mankind simply covers the simple and the naked truth or simply filters out the proper physical information that will let the human brain to simply and to actually balance itself with actual or with true reality or with actual or with true existence or actual creation, which is in constant physical change or in constant physical motion!

In other words, if the human brain simply does not have the proper or the correct physical information or the proper recall of that physical information or the correct point of view, mankind has once again but total or complete chaos!

Actually, there simply cannot really be such a thing as total or complete or full chaos; but almost the exact opposite. It is the simple lack of knowledge or the simple lack of acknowledgement or too much misinformation that causes chaos or that really causes confusion.

Confusion is simply the lack of understanding, limited point of view or chaos is simply the lack of physical matter or simply too much matter or information pressed together, which is simply knowledge.

Confusion is facing a simple mud or brick wall and not being able physically to simply step back a few steps and not being able to simply understand or not being able to simply and physically see beyond the physical mud or the physical brick wall!

That is to simply say, total, full or complete chaos is too much lack of physical or too much lack of solid information or simply too much misunderstanding or too much confusion simply happening at once or too much lack or too much misunderstanding or too much confusion happening one after the other.

Total, full or complete chaos is simply useless blanks or simple cracks where there should be none.

In short, the simple reality is that the human race does not let the human brain to simply complete physical existence or to simply complete physical reality or to simply complete or continue simple creation!

The human brain is simply a reality or existence maker, and the human or mankind does not let the human brain do so. And that is really the simple but the naked or the complete truth!

When the human brain, even a simple animal brain, does not get correct or useable physical feedback through the different senses, the human brain will shut off or will shut down the misinforming senses!

The human brain will even go as far as shutting off physical parts or even shutting off the whole human conscious mind or the whole physical body or both to simply stay in simple physical tune or in simple physical balance with the physical universe or with physical existence or with physical or actual reality or with creation herself!

Sadly, that simple shut down of different or many senses, body parts or simple shut down of the human conscious mind or conscious thought goes on or continues on into many future generations!

Some people are born blind in one eye or both, and parts of their conscious minds, rational or irrational, as well as physical parts from their bodies, do not function or do not work simply because of the lack of brain information or the simple misunderstanding or the simple confusion or limited point view or chaos during their own generations or their past generations!

Sadly, however, modern or even future medicine simply will not kick start that which the human brain simply has shut off!

In other words, human science or human medicine or human technology simply and really cannot deceive or really cannot trick the human brain because human science cannot understand physical existence or physical reality or creation!

But once human science or the conscious living being understands physical existence or actual reality or creation, there will be no need to try to deceive, no need to try to lie or no need to try to trick the human brain into more physical freedom and into more mental or conscious freedom or into more conscious thought.

Many thousands of years ago, the human brain felt safe enough with the very simple physical information, point of view or with the very simple physical knowledge of physical reality or actual existence or actual creation at hand and gave liberty or released the human conscious mind or conscious thought.

Sadly, however, that simple release or that simple freedom of the human conscious mind or conscious thought created the thing called the Pandora Box or more like the Pandora Jar because the interpreter or translator misinterpreted or mistranslated the box for a jar!

That simple release or that simple freedom of the human conscious mind or conscious thought by the human brain simply created but total or a complete chaos, a total or a complete chaos that still goes on today and every day, and will still go on for some time!

That is, until mankind does something about the human conscious chaos, limited point of view or the human conscious confusion or the human conscious condition that mankind simply bestowed upon himself and bestowed upon many future generations!

The simple and the single purpose of the human conscious mind or reasoning or human point of view or conscious thought, therefore, is to simply gather or is to simply bring about more and complete or true physical knowledge of reality or actual existence or even physical creation; and that is the very simple reality or the very simple physical and the very simple naked truth!

That is to say, true or naked knowledge of physical reality or actual existence or physical creation brings truer or more naked knowledge of physical reality or actual existence or physical creation, especially when there is a very simple personal or a very simple mental or conscious interest or point of view and thus allowing for more personal enjoyment of true or complete or naked knowledge!

Instead of giving mankind or the human race simply more human conscious mind or more physical room for conscious reasoning or better point of view, the human brain has simply taken away some parts or taken away some control of the human conscious mind or conscious thought simply because the human brain feels very threatened or feels very insecure by the very freedom or the very liberty of the human conscious mind or the human conscious thought or point of view!

So instead of physically walking erect with the two legs straight in solid ground and with the conscious mind high in the clean air, mankind is physically bending to soon crawl in mud on four legs once again.

That is to simply say, if mankind still has moving body parts; but if not, mankind can physically drag like a simple worm in the simple bed of dust or bed of mud or bed of disgorged matter or bed of disgorged knowledge mankind is also and purposely creating. No conscious living being willingly eats vomit or willingly eats digested matter. The same rule follows for the human brain.

Not until true or real knowledge of physical existence or actual reality or creation personally hits the conscious living being, knowledge of physical existence or actual reality or the naked truth will not be appreciated or will not be fruitful and thus simply will not be enjoyed. Knowledge of physical existence or actual reality or creation is simply for the simple benefit of every conscious living being.

Knowledge of physical existence or knowledge of actual reality or of physical creation is like a physical apple. Every conscious living being simply recognizes the very simple physical apple or a simple fruit, but every conscious living being appreciates more the very simple physical apple or the simple fruit when the very simple physical apple or the simple fruit has being personally tasted and has physically expanded his stomach. The very simple physical expansion of the stomach, therefore, made the very simple physical apple sweeter!

In other words, the very simple physical expansion of the human conscious mind or a better point of view through conscious thought really makes physical existence or creation but much sweeter!

So, therefore, the conscious living being has to personally get involved with true or with complete physical knowledge of existence or of actual reality or of creation even though that complete or that true physical knowledge of existence or of actual reality or of creation came from other conscious living beings.

How long does the human race, therefore, has to wait before the human brain feels physically safe once again, not only to release those parts already taken away from the human conscious mind, but feel physically safe enough to release new and more mind parts or more conscious mind freedom or allows no new point of view?

Does the human race have to wait another three to four thousand years or even much more to feel physically save or secure?

Unfortunately, the human individual does not live that long; and even if the human individual did live that long, what a waste of time!

But, fortunately, the human race does have something very unique and very simple which can penetrate mountains, which can penetrate borders or even can penetrate time. It can even travel faster than the speed of light. It is simply called conscious thought or conscious reason or reasoning or even acknowledgement!

But conscious thought or conscious reason, as is the human conscious mind, is also a very heavy burden to the human brain!

Instead of that conscious thought or that conscious mind simply penetrating through solid rocks or simply penetrating through solid barriers or simply filling the missing parts or the missing pieces of knowledge or filling the gaps of information of physical reality faster than the speed of light, the human race simply becomes very heavy burdened or too heavy and too stiffened and simply cannot physically move and thus simply cannot see or simply cannot walk in the simple light and simply but enjoy!

That is to say, the human race uses a lot of energy and uses a lot of valuable resources or uses simple knowledge or simple information to simply keep but still!

In fact, the human race lives with old or obsolete or useless knowledge or useless physical information of physical reality or actual existence!

Instead of truly increasing the old knowledge of physical reality or physical creation with the new but true knowledge of physical reality or physical creation, no real room is really or is actually left for the new or for any real improvement or advancement of the conscious living being or even the advancement of human history!

Any personal belief or any personal point of view, any personal thought or any personal habit which does not bring more or true or complete knowledge of actual or physical reality or of actual or physical existence or of physical creation is nothing but simply a very heavy burden.

And that which is a heavy burden, a simple burden or not, has no real or has no true purpose.

In other words, that which is a burden is simply nothing but useless and a chaos maker or a reality destroyer.

Human reasoning or human understanding or simple point of view or scientific knowledge of existence or actual reality or physical creation thus far proves exactly the physical opposite of what is simply and actually expected from that very knowledge of existence or actual reality or physical creation or from human simple understanding!

Scientific knowledge of existence or actual reality or physical creation or human science really or actually has the very simple tendency or has the simple bad habit, for any habit is bad and useless, to really prove with what already is or with what already was with that that is but not. But it simply and actually gets much worst!

Countless scientists or countless researchers entice multinational governments into spending enormous amounts of monies, enormous amounts of energies and even enormous amounts of irreplaceable natural resources to simply prove time after time the very exact physical opposite of what the countless scientists or the countless researchers actually or really wanted or physically expected.

Ironically, countless scientists or even countless researchers actually get paid very well for misinformation, for chaos, or simply for false or useless knowledge that simply bankrupt multinational governments or simply bankrupt global economy!

This is truly a lot of physical nothing for a lot of physical something!

That is the simple reason that the conscious living being simply cannot allow a wrongly used tool, or allow an idea or limited point of view that does not bring personal joy or does not bring a personal expansion of conscious mind, simply dictate conscious thought!

No star burns through the physical energy of another. No apple is physically tasted in another mouth!

What, therefore, is the simple purpose of conscious thought or the simple purpose of conscious reason or reasoning?

What, therefore, is the true or the real purpose of conscious effort or what is the real or the true purpose of conscious thought?

The single and the only true purpose of conscious thought or the single and the only true purpose of conscious reason or reasoning is to simply bring full or complete enjoyment or bring enlightenment of true or complete knowledge of existence or actual reality or physical creation, therefore bringing about more true knowledge of existence or actual reality or physical creation, and thus more freedom or more liberty of the human conscious mind!

In other words, the single and simple true purpose of conscious thought or of conscious reason is to simply bring to the human brain full or complete knowledge, full or complete knowledge here and now of human actual or physical surroundings and human personally interacting with the actual or with the physical human surroundings that would otherwise take countless of thousands of physical years to simply gather or to simply find or to even simply understand or to use and thus simply enjoy.

Joy or real enjoyment is actually or is really an expansion of the human conscious mind, but because the joy or the enjoyment, or even a simple conscious thought, is used as energy the expansion does not last and the human conscious mind returns back to normal or to the original surroundings or original state.

Often, therefore, a simple thought is lost as energy through excitement, joy or enjoyment even before that simple thought was completed or acted out.

If the human being or the human race is therefore feeling a bit comfortable, therefore a bit of more freedom of human conscious mind; but if the human being or the human race is simply feeling a bit uncomfortable or simply feeling a bit threatened, therefore less freedom of conscious mind or less freedom of conscious thought or less freedom of conscious reason, and even less physical freedom or less physical movement of body!

The human race is practically going back to the Stone Age, but not before the human race physically turns to stone!

Too often, therefore, a conscious thought is simply eliminated, rejected or simply forgotten or even ignored before the conscious thought is completed or fully understood or fully and simply enjoyed, creating confusion, misunderstanding or chaos to the human brain. Thus, the dream or the nightmare!

A dream or a nightmare is really or actually is a very simple continuation of an incomplete conscious thought or a simple continuation of an incomplete conscious emotion or half-truths or even a new point of view.

Most dreams or most nightmares are really created by the human brain simply trying to understand and to simply enjoy a new conscious thought, a new feeling or a new emotion or a new point of view that really or actually happened but was not completed or was not understood and thus was not fully enjoyed.

And since the human brain really has no previous or true or complete knowledge of that new reality or new creation or new point of view or the brain really has no real information to that new conscious thought, has no complete or real feeling or has no complete or real emotion, the human brain simply searches or simply rewinds the memory of the brain. Thus, the dream or the nightmare!

But that simple dream or that simple search turns into a lasting nightmare or a lasting mental chaos when the human brain simply tries to attach a simple and complete feeling or a simple and complete emotion from memory, but cannot find one.

And so the lasting nightmare or the mental chaos goes on and on and on until the human brain is fully satisfied or simply but stone dead!

The dream or the nightmare or the mental chaos is, therefore, the simple tool, the pipe or the monkey wrench deliberately thrown by the human conscious mind into the gear box of existence that really is the human brain!

Other dreams or other nightmares or other mental chaos is simply caused by physical pain or physical discomfort.

For example, the very simple need for the human body to relieve or to empty the bladder while at sleep can really cause multiple dreams or nightmares!

Even a fever while at sleep can cause multiple dreams or even nightmares.

However, most dreams or most nightmares are simply and actually caused because of misinformation, false knowledge or half-truths or simply because of the need of the human brain to simply know or to simply complete knowledge of physical existence or actual reality or physical creation.

And the reality of the matter is that when the conscious living being has a personal understanding or has really learned something of reality, reality is no longer the same!

By the way, depending on what side of the body that is slept on, thus the dream or thus the nightmare or thus the mental chaos!

Sleep on the left side of the body, one type of dream or one type of nightmare. Sleep on the right side of the body, another type of dream or another type of nightmare.

The dream is as sweet or as bitter as the physical side of the body that is simply slept on.

The very simple reason for having different types of dreams or having different types of nightmares depends on what side of the body that is slept on is that the right side of the human conscious mind or the right of the human brain physically controls the left side of the body and the left side of the human conscious mind or the left side of the human brain physically controls the right side of the body.

In other words, the right side of the human conscious mind or the right side of the human brain, which is irrational or which is emotional, simply looks for information or true or real knowledge dealing with the emotions.

If no information, no recollection or no knowledge or no good or no meaningful feeling or no real emotion is found in memory, then the simple nightmare.

The nightmare or the chaos is the simple tool to try to explain an incomplete or a misunderstood human emotion or a misunderstood human feeling. The dream or the nightmare is the simple physical motion or point of view to simply try to understand an incomplete thought or an incomplete emotion or even a new experience or the reaction to that new experience.

Actually, a dream or a nightmare or a mental chaos is simply caused by a very simple denial or by half-truths or by false knowledge of physical existence or by false information. Even a simple lie can cause a lasting dream or a lasting nightmare or even a lasting mental chaos.

The left side of the human conscious mind or the left side of the human brain, which is rational or logical, looks also for complete information dealing with physical reality or with complete truths. If the opposite is true or is confusing, then the dreams or the lasting nightmares.

The dream or the lasting nightmare here is supposed to explain the irrational that happened in the rational or the rational in the irrational.

For example, the rational or the logical side of the human conscious mind or the human brain is simply trying to understand and trying to enjoy a physical action or point of view that really or actually happened, such as a real explosion, but the individual or the person consciously denied that physical or that real action or point of view through conscious disbelieve or not really believing that such a physical reality actually or really happened even though he personally saw the actual physical explosion, therefore really causing confusion or really putting the human brain in total chaos!

Another very interesting thing about dreams or nightmares or even mental chaos is that they are too often become multiple dreams or multiple nightmares or even multiple mental chaos in a single dream or a single nightmare or even a single mental chaos.

That is, many dreams or many nightmares in one single dream.

Actually, a three dimensional dream or nightmare!

To the dreamer it was one continuous dream. But in reality, they were multiple dreams, at times more irrational than rational.

It is more like a dimensional dream or many points of view, where time stands still but many things are really or are actually happening at once and at the very same space or location!

This dimensional dream is the brain really searching or really analyzing multiple dreams or multiple points of view at the very same time!

The brain is really searching for a true or a real reason as to what made the conscious living being act or react irrational in the rational or act or react rational in the irrational.

In other words, since the brain cannot understand the new or the incomplete conscious thought or the new point of view or even a partial reality or a partial emotion or partial feeling, the brain looks or rewinds the memory in order to complete that new conscious thought, that new point of view or that new but irrational emotion or partial feeling or even a partial reality.

And so the lasting dream or the lasting nightmare really is equal to the reality acknowledged or to the reality already recognized plus the point of view or the simple understanding of that recognized reality minus the reality unseen or unrecognized or even the actual reality denied.

In short, "D = +0-." Or simply put, plus zero minus.

Once again, the simple and the single true or real purpose of the human conscious mind is to simply but know and to simply but enjoy true or complete knowledge or to simply enjoy actual or physical reality or physical creation.

The true or the real purpose of conscious thought or of conscious reason or a much better point of view is simply to complete knowledge or to simply complete information of actual existence or actual reality or physical creation, here and now, and thus give more freedom or more liberty to the human conscious mind!

Not until the human conscious mind, as well as the human brain, is really complete or fully satisfied with the actual or the physical reality or the physical creation that really is here and now, there will be no more mental freedom or no more liberty of human conscious thought or no more conscious mind and no more liberty or no more physical movement of human body!

That is to simply say, not until the human conscious mind or the human brain is really complete or filled or fully satisfied with actual or with physical reality, there will not be more human conscious mind, no more knowledge or no more understanding or no new point of view and thus no more enjoyment!

That is the simple reason that human or mankind, as well as the human individual no matter how great or how smart, gets stuck and human civilization, as well as the human individual, crumbles to useless pieces time after time.

Knowing, simply understanding, and really enjoying reality, physical and actual existence or even physical or truth knowledge brings about an expansion of the human conscious mind and more freedom or more liberty of physical body.

And that is the simple reality or the simple and thus the naked and the complete simple truth!

There is no real or true point of knowing, understanding or having complete or having full knowledge of existence or actual reality or physical creation if nothing is really going to be done with that complete or perfect knowledge of reality!

True or perfect knowledge or true or real information is to know that "one simply but are!"

That is to really say, that one can simply be but more.

True or real knowledge of reality or creation or true or real information is to know that one simply exists and that one can really do something more about it!

To stand still or to stand motionless, therefore, is to have simply but existed and thus simply and actually just waiting to simply die once again, only this time through a lot of physical pain!

Wouldn't it be more than simply and wonderfully grandiose to simply know that mankind, really the conscious living being, but simply is the simple and the only actual answer or the simple and the only true or real solution to the final or to the ultimate question, to the final or to the ultimate problem or to the final or to the ultimate solution that has yet to happen?

Existence or physical creation created naked man but not mankind or not the human being. Mankind or the human living being consciously created or consciously completed mankind or the human conscious being when naked man physically broke all the laws of Nature!

The simple and the single true or real purpose, therefore, of the human brain is simply to complete or to make perfect existence, to complete or to make perfect physical reality or to complete or to make perfect physical creation or to complete the gaps of true or real knowledge!

That is, to acknowledge or to renew existence!

The simple and the single true or real purpose of the human conscious mind is simply the very same as the very simple and the single true or real purpose of the human brain, but with full understanding or with complete personal enjoyment!

That is really to say, the simple true or real purpose of the human conscious mind, through conscious thought or through conscious effort, is simply to help the human brain here and now to simply finish or to simply complete or to simply make perfect physical existence or complete or make perfect physical reality or complete or make perfect physical creation in and through personal enjoyment or in and through personal enlightenment or through personal acknowledgement!

So, it is a very simple question of how conscious living beings through conscious thought or through conscious effort and enjoyment will finish or will complete physical creation or finish or complete physical Existence!

By the way, true, real or complete knowledge is that knowledge that can physically be but proven because true knowledge simply exists!

Chapter One

Why is the universe round?

Why is the universe or creation round? Or why is the universe or why is creation sphere-like?

The universe or creation is simply round because of why!

But what is why? Why "why?"

What makes why?

The question, therefore, is the physical foundation or is the point of view that will be built from or that will be built on!

The using of "why?" as a physical foundation, as a point of view, as a physical situation or as a physical or as an actual position to really built from or built on is the simple reason that the very childish, the very foolish and thus the unanswerable or foundationless question cannot be understood and in order to and understand the question, the situation or the physical position and find the real or the true answer, one must simply change the very childish, the very foolish and thus the unanswerable, foundationless and thus the meaningless question around!

In other words, one must change that childish point of view. When a question is asked, one has put one self in a situation or in a physical position or in a higher point of view.

The simpler the situation, the simpler the foundation or the simpler the physical position or point of view, therefore, the simpler the answer will be.

No one really starts a real or an actual quest or really starts a real physical journey without a real or an actual reason or a real or an actual purpose or a real physical position or a better point of view.

And that simple purpose or that physical position or better point of view is for the simple enjoyment or simply for conscious enlightenment or complete conscious joy!

Conscious enlightenment or complete conscious gladness and joy in reality is an expansion of conscious mind.

However, there cannot really be any conscious gladness or any conscious enlightenment or any complete conscious joy unless there is really an expansion of conscious mind or a better point of view through real, true, useful, meaningful or complete knowledge of physical reality or actual existence or physical creation.

That is to say, there cannot be an expansion of conscious mind or a better point of view without complete conscious thought or complete awareness.

Actually, there really cannot be any conscious enlightenment or a better point of view or there really cannot be an expansion of conscious mind without actual, real or satisfying knowledge or without conscious thought or without acknowledgement!

That is to say, there is no expansion of conscious mind or there is no conscious enlightenment without the personal knowledge of actual reality or without the personal knowledge of actual existence or physical creation.

Actual existence or actual knowledge of existence or of creation brings joy or conscious gladness or brings an expansion of conscious mind. To be very joyful, to be very glad or to be enlightened, therefore, requires but complete knowledge of physical reality, physical creation, or actual existence or complete conscious thought, which is really complete or full awareness.

In other word, understanding and thus fully enjoying the "I am" factor or fully understanding and thus fully enjoying the "I simply exist" and I can really enjoy doing something or really enjoy doing more about it!

No one, not even an eccentric or a being with a higher conscious mode or mind, really starts or begins an equation or starts or simply begins an addition or subtraction problem without first personally understanding the simple workings of the equation itself or without first personally understanding something of addition.

It is the very personal understanding, the very personal recognition, and the very personal recollection or the very personal reposition or the complete knowledge of actual existence, reality or creation that will really bring more useful knowledge of reality or actual existence or creation and therefore more conscious enjoyment or more conscious enlightenment.

Enlightenment or complete joy or complete conscious thought is a simple expansion of conscious mind. Such an enlightenment or complete joy causes a mental high or a mental ecstasy!

Actually, the simple expansion of the human conscious mind through true, real, satisfying or complete or useful knowledge brings forth complete or true joy or total conscious awareness or simply conscious enlightenment or acknowledgement.

Ironically, some knowledge, though true and thus very useful, is very useful only to some conscious living beings but not useful to other conscious living beings.

And although that true or real knowledge is very useful, that knowledge or that information does not bring joy or enlightenment or expansion of conscious mind or better point of view to those who have such useful knowledge or possess such useful knowledge!

However, there will be a point in time where all true or all real knowledge of existence or complete reality or physical creation or all useful physical information will be used by every conscious living being for the further full benefit of all conscious living beings!

The more mankind personally knows or the more mankind personally learns about the way existence or physical creation really or truly functions or the way reality works, the better or the more independent the human race will be.

That is to say, the more true knowledge of existence or of actual reality or of physical creation a human individual or a conscious living being personally has about existence, the better or the merrier the human race or the other conscious living beings will be. It will be like a big bang of knowledge of reality or a big bang of conscious enlightenment or acknowledgement!

In other words, the conscious mind will be high in real or be in complete knowledge of existence or actual reality or physical creation.

But such a conscious high or such a conscious enlightenment or expansion of conscious mind or complete joy will not be a one-time deal. The conscious high, the conscious joy, the conscious knowledge of reality or conscious enlightenment will be permanent!

The simple question or the simple quest, therefore, is only as good as the point of view and as good as the true anticipation of the answer!

Q equals the reality at hand plus the point of view minus the irrational or minus the reality not yet seen!

Or simply put, "Q = plus zero minus," where plus equals reality; zero equals point of view; and minus equals the reality not yet seen!

The greater the irrational or the greater the exaggeration or the greater the lack of knowledge of unseen reality, the less the point of view or the lesser the understanding, and, therefore, the less the physical or the actual reality or physical creation!

Interestingly, even though knowledge may be lacking, simply getting interested, getting glad or even excited about the upcoming answer really improves point of view or makes room for further understanding, and may actually lead to a true or to a real answer; and therefore, a greater reality!

A very good example of "Q = Plus Zero Minus" is as follows. For example, a drinking glass or a water pitcher as plus zero minus. If the drinking glass or the water pitcher is full, most observers will agree or will have the same point of view.

If the drinking glass or the water pitcher is empty, most observers will also agree or will also have the same point of view.

But if the drinking glass or the water pitcher is half full, then most observers will have different point of views. Some observers will say that the glass is half empty; and some other observers will say that the glass is half full.

There will even be those few observers that would argue that the glass is not really completely full or that the glass is less or even more than half, without measuring the glass. Those few observers are those with half-truths, good sounding lies or false science.

Now, getting back to the track!

Very few observers will really say that the glass is really both.

That is, that the glass is half full with water and half full with space. Even an empty glass has space or emptiness; and interestingly, a full glass of water lacks space.

The simple question to the origin of the physical universe, to the physical origin of creation or to the origin existence, therefore, must be simply asked: What makes the universe round?

And thus, there it is part of the answer!

Now, in order to find the rest of the very simple and the meaningful answer one must first start or must first begin by asking many more meaningful and answerable questions, for in order to add, one needs but a few.

True or naked knowledge of existence or actual reality or physical creation really brings forth but more true, naked or complete knowledge of existence or actual reality or physical creation or acknowledgement or even enlightenment!

That is to say, the simple true or the naked answer is actually or really found in many more simple and real or answerable questions; add the simple and the real and the answerable questions and one will get the simple and the real and the final and the true and satisfying answer.

A true or a real answer is, therefore, a useful answer. And a useful or a real or a true answer is that simple answer which is complete and in being complete, therefore, brings great joy or brings gladness.

Complete knowledge of existence or of actual reality or of physical creation really brings but complete joy, complete gladness or but complete enlightenment, which simply is an expansion or is a new or better point of view of the conscious mind.

What, therefore, makes the universe round or sphere-like?

That is to simply say, what makes it possible for the universe to be round or sphere-like?

To simply answer the above, very simple and actual or real question, one has to simply and also ask: "What makes the universe?"

Or "what really makes possible the universe or what makes possible Existence?"

That is to say, what is the entire universe made up of?

In other words, what are the actual or the real things that actually and really make but possible the universe or what makes possible but Existence or physical creation?

What are the actual or the real things that add to or what are the real things that really make Existence or physical creation possible?

What are the recognized physical things that really make possible the universe?

And how these recognized physical or recognized solid things really make the universe round?

The universe or physical Existence or physical creation is made up of space, darkness, and coldness.

The universe or physical reality or physical Existence or creation is also made up of matter, light, and heat.

The universe or part of physical Existence or physical creation is also round or sphere-like, and the universe or part of physical or actual Existence or physical creation is also physically moving or physically expanding throughout the vacuum of space or physically moving throughout the sphere or throughout the round and invisible shield.

Actually, the universe or physical creation is really part of a whole, which is Existence or reality.

In short, the universe or physical creation is simply made up of space, darkness, and coldness.

The universe or physical creation also is made up of physical or solid matter, light, and heat. Physical motion or physical movement is also a major part of the universe or of physical creation.

Physical motion or physical moment is also a major part of actual or of real or of physical Existence or physical creation.

And all of these and real or actual things are really interacting and are physically expanding inside a so-called sphere or in a round and invisible but real field.

What, then, is space?

Space is the physical absence of a physical or a solid matter, light or energy. Space is the absence of a physical or of a solid body or the physical absence of a solid or of a physical identity.

Space also has no mass or space has no density or weight. Space is, therefore, non-existence or nothing!

Space is also a vacuum or emptiness. Space, is therefore, the absence of a physical reality.

That is to say, space is nothing, space is empty and space is lacking. Space is measured, however, not by the amount of space or the amount of emptiness but by the physical size of the actual or the real thing or the physical or the solid matter, light or energy or even heat that is going to actually fit or actually occupy that space or that emptiness.

So in reality, space in the physical sense simply does not exist or space is non-existence!

What is darkness?

Darkness is also the physical absence or the lack of light or Darkness is also the physical absence of energy.

Darkness also has no mass or Darkness has no density.

And what is cold or coldness?

Cold or coldness is also the physical absence or the lack of heat. Cold or coldness is the absence or is the lack energy or heat.

Cold or coldness also has no mass or cold or coldness has no density.

What is matter?

Matter is that which occupies space and matter is also that which has mass.

Matter is that which is physical or is solid. Matter is that that has mass and matter is also that which has density.

Matter is also physically affected by motion or by physical movement or by time.

Matter also physically exists or matter simply but physically is.

Matter, is therefore, physical reality. Matter simply is physical knowledge of existence or actual reality!

That is, matter is also a product of physical reality or physical creation.

In fact, if no matter, then no physical reality, no universe or no physical creation!

What is light?

Light is that which occupies space. Light also has mass and light also has density.

Light also physically exists or light simply but is.

Also, light is a physical product of physical reality or of physical creation.

What is heat?

Heat is that which occupies space. Heat also has mass and heat also has density.

Heat also physically exists or heat simply but is.

Also, heat is a physical product of physical reality or of physical creation.

Thus, darkness is to space as light is to energy or to matter!

That is to say, darkness is to space or to cold or to coldness as matter is to energy or to light!

In short, space is to vacuum or to emptiness, to darkness and to coldness as matter is to motion or to expansion, to energy and to light.

Also, space is to nothing or space is to nothingness as matter is to something or is to physical reality.

In reality, therefore, space is to complete emptiness or space is to unseen reality as matter is to physical fullness or to physical reality.

Now let's simply say once again that space, darkness and coldness are simply but the very same thing.

And let's say also once again that matter, light and heat or energy are simply but the very same thing.

Now, there are two very simple opposites to work with or to work in: Nothing and something, or space and matter, or Non-existence and physical Existence, or Negative and Positive, or Black and White.

Or simply, that which is and that which is but not!

Space, coldness or darkness and matter, energy or light actually or really form or create the universe or physical creation, which is round or sphere-like, where space, coldness or darkness and matter, energy or light simply unite, physically join or simply intersect or physically meet or even divide.

Now, what is the solid or the physical thing with mass or with density that makes but a simple sphere or even a round shield?

What makes but a field or a sphere or even a simple shield?

What two figures or what two objects or what two things make a simple sphere or a simple circle where the two things physically unite, physically meet, physically interact or physically join or even divide?

What really makes but a sphere or shield?

What really makes a round field?

Or what is the very thing that makes a field?

What about two magnets?

What about two magnets in the form of squares or even in the form of cubes?

Yes! Where the two magnets or where the two magnetic cubes or the two magnetic squares, in this case matter and space, physically join, physically react or interact or physically unite or even divide, at the corners, the two magnetic cubes or the two magnets actually or really form a simple sphere or a field or a shield.

In fact, many or countless spheres, fields or shields or many or countless circles are physically formed by the two magnets or by the two infinite magnetic cubes or by the two magnetic squares!

But how are the magnetic spheres or the magnetic circles or the magnetic field or magnetic shield formed?

The magnetic spheres or the magnetic circles are formed due to the magnetic fields or magnetic shields of each of the magnetic cubes or the two magnetic squares.

An excellent or a perfect and the most simple and the most unique example is the field or shield of a bar magnet. See Figure or see Drawing Number One, the Field of a Bar Magnet.

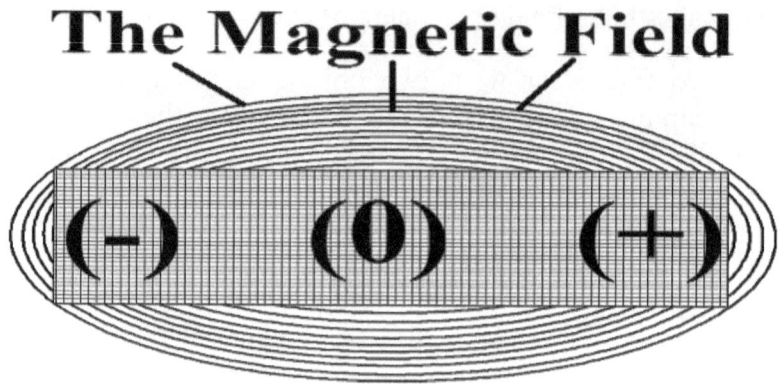

The bar magnet simply creates or simply makes a special physical condition in the empty space or the emptiness around the bar magnet.

That space or that special physical condition around the bar magnet is really noted as a magnetic field or as a magnetic sphere or as a magnetic shield.

The magnetic field or shield can also be simply called a field or shield of physical attraction.

The two cubes' or the two squares' physical or magnetic attraction for the other cube or for the other square makes possible the magnetic spheres or makes possible the magnetic circles.

That is to simply say, instead of the magnetic field or shield or the magnetic sphere physically going around each of the cubes or each of the squares as the magnetic field or shield does with a bar magnet, the magnetic field or the magnetic sphere simply turns toward the opposite magnetic cube or the opposite square, thus making really possible the magnetic field or shield or the magnetic spheres or the magnetic circles or even the magnetic rings.

To simply illustrate or to simply show the magnetic field or shield, the magnetic spheres or the magnetic circles of the two infinite magnets physically attracting or physically interacting with one another, the Grid or the Rectangular Coordinate System is used.

Actually, the Cartesian coordinate system is used as in Figure or as in Drawing Number Two. See Figure Number Two, The Magnetic Spheres, for more clarification.

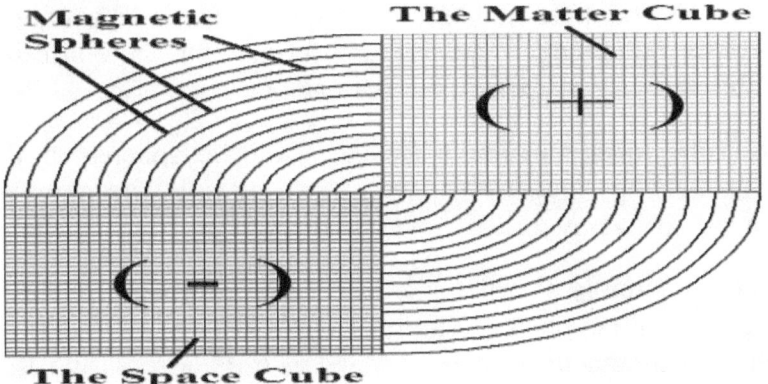

Only two sections or only two quadrants, sections or quadrants number two and sections or quadrants number four of the Grid or the Cartesian coordinate system are used. Quadrant number one and quadrant number three can also be used to show or to illustrate the two simple infinite magnets or the two simple magnetic cubes.

The simple reason for this selection of the two sections or the two quadrants is that the selected two sections or two quadrants physically meet or physically interact only at one physical point, and that physical point really is physical point zero, physical point of impact, physical point of intersection or physical point of interaction or physical point of origin or even physical point of view.

It can also be simply said that the two selected sections or quadrants are physically opposite to one another or facing one another.

As it can be simply seen from the simple diagram or the simple drawing, Figure Number Two, where the two magnetic cubes or where the two magnets physically meet or even divide, many magnetic circles or many magnetic spheres are physically formed!

But the magnetic circles or the magnetic rings or the magnetic fields or magnetic shields must be sphere-like, for they are really the product of two magnetic cubes.

In reality, countless numbers of magnetic circles or countless numbers of magnetic spheres are formed for the Cartesian coordinate system goes on forever and ever.

Also, the magnetic circles or the magnetic spheres or the magnetic rings increase in size as real or as whole numbers increase in size by one.

That is to simply say, the size of the first magnetic sphere or magnetic ring is one or is number one. The second magnetic sphere or magnetic ring is two or twice the size as the first magnetic sphere or magnetic ring. And the third magnetic sphere or magnetic ring is three times the size of the first magnetic sphere or the first magnetic ring, and so on and so on.

Now, inside one of those magnetic spheres or inside one of those magnetic circles really is the universe!

When the two magnets, in this case the space cube and the matter cube or the light cube, physically attract one another through their magnetic fields, the two magnetic cubes create many magnetic spheres.

Inside one of those magnetic spheres is the universe. See Figure or Drawing Number Three, the universe.

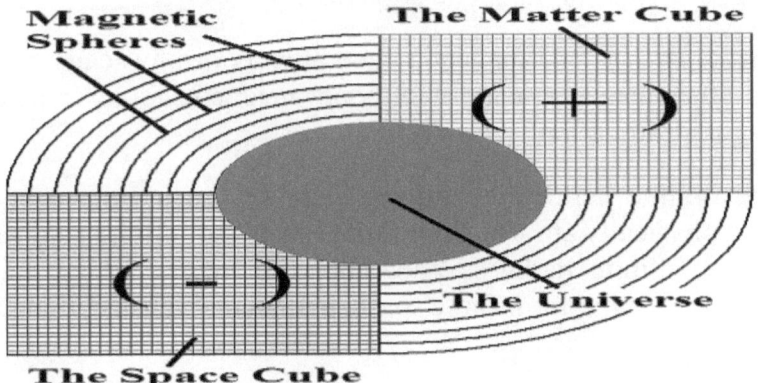

But what, therefore, makes possible the magnetic cube?

What makes possible the cube?

That is to say, how does the cube gets its simple and its unique shape?

Since the infinite magnetic cube is outside of space or is outside any vacuum or is outside any gravity or is outside any compression, there really is no compression or is no vacuum to physically act or to physically interact upon the infinite magnetic cube, thus allowing the infinite magnetic cube to simply expand throughout and, therefore, making or taking its unique and simple cube shape.

Now there are two infinite magnetic cubes and, where they physically meet, physically interact or physically intersect, countless magnetic spheres or countless magnetic circles are formed; thus, the physical universe.

One of those magnetic spheres or one of those magnetic circles is the physical universe. See Figure Number Three, the Universe.

Actually, every magnetic sphere or every magnetic circle is really a separate universe or a separate dimension of time!

Both infinite magnetic cubes physically attract one another or one cube physically attracts the other cube with the very same magnetic force or magnetic attraction.

That is to say, one magnetic cube attracts or physically pulls the other magnetic cube with equal magnetic force or equal magnetic attraction.

The infinite matter, the infinite light or the infinite energy cube physically attracts the infinite space, the infinite coldness or the infinite darkness cube, but the infinite space cube releases nothing for infinite space, being empty, has nothing to release.

The infinite space, the infinite coldness or the infinite darkness cube physically attracts the infinite matter, the infinite energy or the infinite light cube, which has density or is solid or is physical, and a physical part of the infinite matter, the infinite energy or the infinite light cube is broken off or is released, and takes up space in the magnetic sphere or takes up space in the magnetic circle formed in the vacuum of space by the two infinite magnetic cubes. See Figure Number Four, Matter in Space.

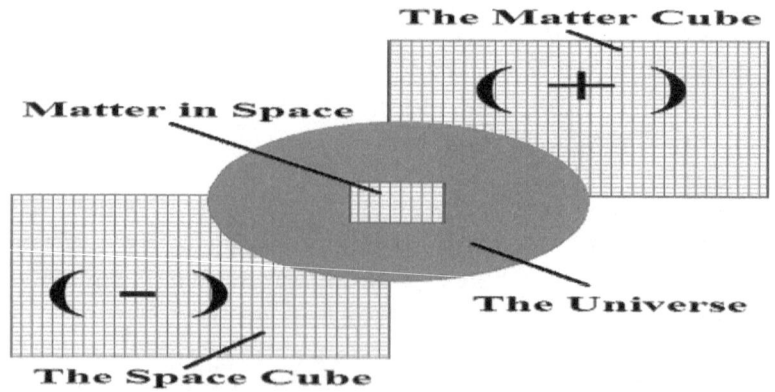

Once the solid or the solid part or the solid piece of matter which is in the physical form of a smaller cube of matter, or simply call it the Universe's solid egg, from the infinite matter, the infinite energy or the infinite light cube, physically enters the vacuum or the compression of space, there is the actual or the real and the physical beginning or the physical laying of the Universe's egg.

Since empty space is super cold and also a super or a giant vacuum, the physical properties of matter, energy or light, or the smaller matter, energy or light cube as an egg in the vacuum of space, physically change.

Matter, energy or light now in the super vacuum or in the super compression of space shrinks or matter, energy or light is super compressed in the super vacuum of space.

Once super compressed, matter, energy or light as a smaller cube is now super compressed into a sphere of matter, energy or light.

The new sphere of matter, energy or light in the super vacuum of space super heats and super explodes and thus simply expands throughout the universe or simply expands throughout the simple magnetic sphere or simply expands throughout the magnetic circle or magnetic field or magnetic shield.

Thus, the Big Bang!

Actually, the Big Bangs! See Figure or Drawing Number Five, the Expanding Universe.

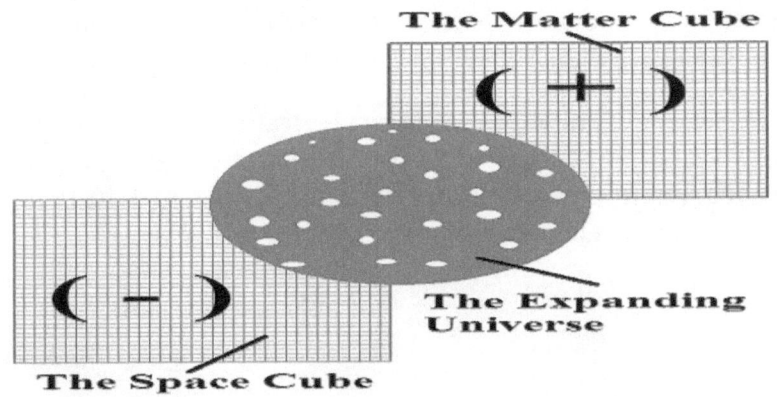

As it can be really be seen, the rest of the universe or the rest of actual or real existence or even creation is time, physical evolution, history and simple ideas, true or not.

That is to simply say, the rest of the real or the actual universe or the rest of the real or actual reality or physical existence or physical creation is simply the Big Bangs of human ideas!

The reality of the matter is that there are really two creations. There is physical creation or physical reality. And there is mental creation, but mental creation or human thought does not let physical creation or actual reality thus be seen, and much less enjoyed. Just because it is believed, for example, it does not make it so. The truth or reality is not something one just believes in, but the truth or reality is something that can really be seen. When mankind does not understand something or even see it, mankind humanizes it or makes it unnatural.

The universe or existence or physical creation is as the conscious living being simply is. The more the conscious living being simply is or the more the conscious living being simply becomes or really knows or understands, the more the universe or the more physical or the more actual existence or the more physical reality or the more physical creation simply will be!

In other words, the universe or existence or even physical creation is only limited to the conscious thought or only limited to the conscious belief of the conscious living being. Limit the conscious knowledge of the universe or limit the knowledge of actual existence or of actual physical reality or of physical creation, thus limit the conscious living being.

Existence, the universe or physical creation is composed or really made up of three main parts. The three main parts are physical reality, neutral reality or point of view, and the reality not seen but can be negative or irrational.

Or simply put, all of Existence is "(+0-)."

In other words, E = (+0-.) That is, plus zero minus. The same formula can also be used for the universe.

That is, "U = (+0-.)" Thus, the universe is really equal to the actual reality recognized plus personal point of view or personal understanding or the reality actually seen minus the reality not yet recognized or the reality believed.

The greater the plus is the greater the point of view, and the lesser the negative or the lesser the irrational.

The greater the negative or the more the irrational, the less the point of view or the less of actual physical creation is really seen and, therefore, the less understood and the less enjoyed or the less acknowledged.

But if the understanding or if the enjoyment or if the acknowledgement is greater, thus greater is the universe or thus greater is existence.

Chapter Two

Dimensions of time

In reality, there is a finite or there are only 118 separate universes. In reality, there is a finite or there are only 118 dimensions of time between the two infinite magnetic cubes: the space, the cold or the darkness cube, and the matter, the energy or the light cube.

Even though the two infinite cubes are encircled in magnetic spheres, the first 118 magnetic spheres or the first 118 magnetic circles are actual universes or are actual dimension of times. The first 118 magnetic spheres are actually the middle, the center or the belly bottom of existence or of reality or physical creation.

Every magnetic sphere or every magnetic circle in Figure or in Drawing Number Two, for example, simply illustrates a separate universe or a separate dimension of time.

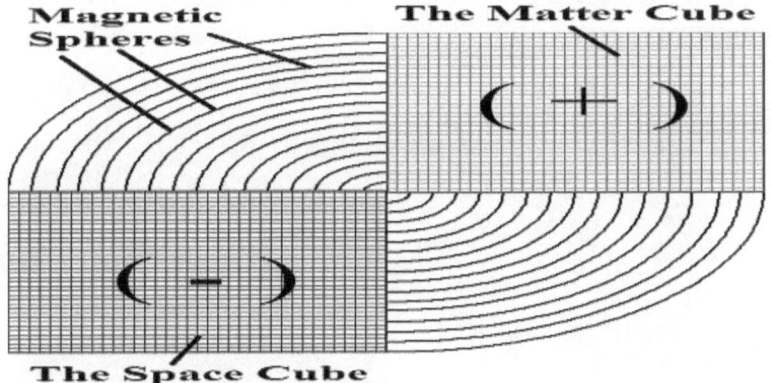

In Figure Number Two, for example, there are fourteen circles representing or simply illustrating fourteen magnetic spheres. Figure Number Two is actually or really showing fourteen separate universes or really showing fourteen separate dimensions of time.

The number of universes or the number of dimensions of time, once again, is in reality finite or the number is limit even though the magnetic spheres or the magnetic circles are infinite or countless as the numbers are infinite or countless.

What this actually or what this really means, is a magnetic sphere or a magnetic circle inside a magnetic sphere inside a magnetic sphere until infinity.

However, only the first 118 magnetic spheres or only the first 118 magnetic circles are actual universes or are actual dimensions of time. Each magnetic sphere or every universe is separated by a dimension of time or is separated by a dimension of space.

The magnetic circles or the magnetic spheres increase in size as the two infinite magnetic cubes extend forever or extend until infinity.

The physical motion, the physical movement or the physical expansion in Time or in the universe or the physical expansion of the universes is simply created or is simply caused by the magnetic field of the two infinite magnetic cubes. See Figure Number Six, Space-time.

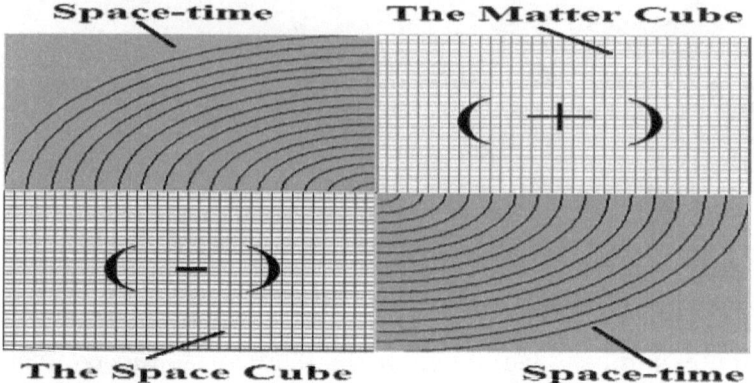

In Figure Number six, notice that there are now two more infinite cubes. Now then, those two extra infinite cubes are the Time cubes or are the space-time cubes. Although those two extra infinite cubes are not directly responsible for physical Time, most of the magnetic fields or most of the magnetic spheres and the physical expansion or the physical motion of the universe or universes is in or within those two extra and infinite motion or time cubes.

The two extra infinite motion cubes also indirectly add more empty space to the universe and universes. Henceforth, space-time!

The space-time cubes can also be called the Motion Cubes, but only because of the magnetic field that goes through them.

By the way, not every magnetic sphere is a dimension of time and not every magnetic sphere is a separate universe. The dimensions of time are finite as the universes are finite in number. There are only 118 dimensions of time as there are also only 118 separate universes.

When empty space or the vacuum of space physically attracts the infinite matter, the infinite energy or the infinite light cube, not only does the infinite matter, the infinite energy or the infinite light cube releases a smaller matter, energy or light cube with 118 smaller matter, energy or light cubes, but also the infinite matter, the infinite energy or the infinite light cube releases multiple matter, energy or light cubes at the very same time.

The infinite matter, the infinite energy or the infinite light cube releases smaller matter, energy or light cubes with 118 smaller cubes each according to the size of the magnetic spheres or the size of the universes.

That is to say, the infinite matter, the infinite energy or the infinite light cube releases at the very same time one matter, one energy or one light cube with 118 pieces of matter, energy or light for the number one sphere or the number one or the very first universe; the infinite matter, the infinite energy or the infinite light cube releases two matter, energy or light cubes with 236 pieces of matter, energy or light for the number two or the second sphere or the number two or the second universe; the infinite matter, the infinite energy or the infinite light cube releases three matter, energy or light cubes with 354 pieces of matter, energy of light for the number three or the third sphere or releases three matter, energy or light cubes for the number three or the third universe, and so on until the very last universe or on until the very last dimension of time, which is dimension number 118. Dimension number 118 or universe number 118 is 118 times the size of dimension number one or 118 times the size of universe number one.

In other words, magnetic sphere number 118 is 118 times the size of magnetic sphere number one. Universe number 118 is really 118 times the size of universe number one.

For every one cube of pure matter, energy or light, there are 118 smaller matter, energy or light cubes or 118 pieces of matter making that one matter, energy or light cube.

In the simple case of Figure Number Six, fourteen different or fourteen separate universes or fourteen dimensions of time actually began at once or really began spontaneously.

The very same number of matter, energy or light that physically enters the vacuum of space is the very same number of space taken or physically occupied by the matter, the energy or the light.

As soon as matter, energy or light from the infinite matter, the infinite energy or the infinite light cube physically enters empty space, cold or darkness or matter enters the vacuum of space or enters the magnetic spheres created by the magnetic field of both the infinite matter cube and the infinite space cube, matter, energy or light in the vacuum of space is super compressed due to the enormous pressure or due to the enormous vacuum of space, cold or darkness.

The enormous pressure or the enormous vacuum in space causes matter, energy or light in the vacuum of space to super compress into a giant or into an enormous magnetic sphere of matter, energy or light in the vacuum of space and to superheat and to super explode, thus making possible the Big Bang!

Actually, making possible the Big Bangs! Or the hatching or the hatchings!

Enormous energy is loss from the super Big Bangs! Or from the super hatchings! But a lot of physical or solid matter, energy or light is still left in the vacuum of space to spread or to expand throughout the vacuum of space and spread or expand throughout the universes or throughout the magnetic spheres.

Some matter will lose energy, therefore, becoming stars, planets and moons and some matter actually becoming even smaller pieces or bits of matter, such as space rocks or space dust.

The heaviest or the denser matter or the heaviest or the denser stars at one point in time or at one point in the vacuum of space will turn into supernovas and then into super black holes or turn into the cancer of the universe and universes.

Actually, black holes or collapsed stars are like giant vacuum cleaners or giant space sweepers, really giant magnetic graves!

There will be another point in time or another point in the vacuum of space when most of the universe and universes will be occupied by super or by giant black holes or by giant cancerous stars or dark stars or vast empty magnetic graves.

Once again, the universe or universes will be occupied by giant vacuum cleaners or giant sweepers or giant graves.

The universe and universes will look like a giant Swiss cheese, but without the appetizing smell. See Figure Number Seven, the Imploding Universe.

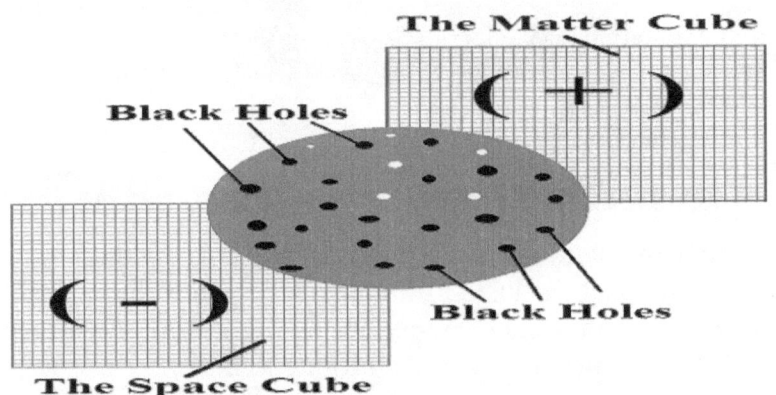

At that very stage, when the entire universe and all universes are occupied by enormous black holes or enormous dark stars or enormous vacuum clears, the enormous black holes or the enormous dark stars will disintegrate or will drain the entire light or the entire energy or the entire matter of the universe and universes spontaneously or at once!

That is to say, the giant black holes or the giant space vacuum clears will begin contracting or moving or going to the center of the vacuum of space or to the actual physical point of impact or to point zero or to the point of origin until the entire physical universe and the other universes collapse—the exact opposite of the Big Bang! Or Big Bangs!

The black holes or the collapsed stars will begin to move toward the center or begin moving toward the actual physical beginning or the physical origin of the Universe and universes, removing all the matter, all the energy or all the light from the entire universe and universes and thus making one giant or making one enormous black hole or one giant or one enormous vacuum per universe or one enormous vacuum per magnetic sphere. See Figure Number Eight, Point Zero or Point of Impact.

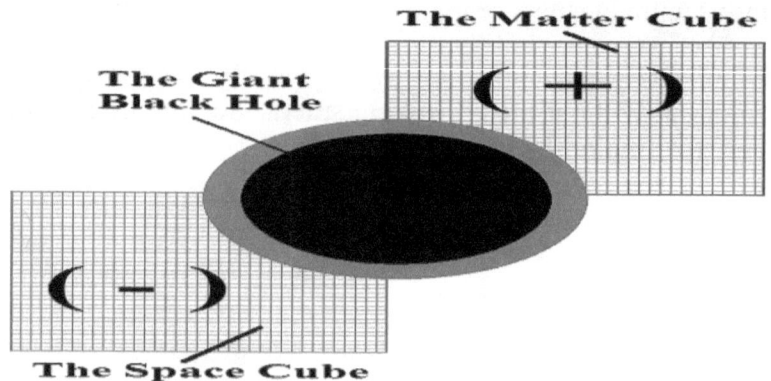

It is this new giant or this enormous black hole vacuum in the vacuum of space or inside the magnetic sphere that actually attracts new matter, new energy or new light from the infinite matter, the infinite energy or the infinite light cube.

Once new matter, new energy or new light from the infinite matter, the infinite energy or the infinite light cube enters or re-enters the giant black hole vacuum in the vacuum of space or in the magnetic sphere, the new matter, the new energy or the new light is super compressed recreating the Big Bang! Or really creating the Big Bangs! And destroying the giant or the enormous black hole vacuum!

Actually, every giant or every enormous black hole in the vacuum of space or in every occupied magnetic sphere or in every occupied dimension of time or in every occupied dimension of space will be destroyed at once or at the very same time by new matter from the infinite matter, the infinite energy or the infinite light cube!

In other words, every giant black hole vacuum occupying the 118 universes or occupying the 118 dimensions of time will be destroyed at the very same time by new matter, new energy or new light from the infinite matter, the infinite energy or the infinite light cube.

However, there also exists the possibility that the giant black hole or the black holes will disperse as if a storm and space will become as if a sea of tranquility.

The Stars and the Black Holes

The universe is a neutral point or is a point zero which is converted into a positive point or into a one or more when light or matter enters into the universe. The universe also is converted into a negative point when the grand majority of light turns off or the majority of matter no longer has energy...

Before light or matter enters into the universe the weight of the universe is cero and when light or matter enters into the universe thus the universe takes on weight. Light or matter is composed of 118 Elements and the weight of each element is two times its atomic number, more like its positive number.

In the case of element number one, for example, its weight is of two and in the case of element number two its weight is of four and in the case of element number 118 its weight is of 236...

Curiously, that the totality of the 118 elements adds to one and that the totality of their weight adds to 2.

That is to say, that if we added from one to 118 thus the sum would be 7,021. And if we added that sum thus it would be 10 and if we add that last sum thus it would be one. That is, 118 is equal to one...

And if we added from two to 236 the sum would be 14, 042. And if we added that sum thus it would be 11 and if we added that last sum thus it would be 2. That is, 236 are equal to two...

Thus, light or matter enters into the vacuum of space or into the universe as one or as a unit which is composed of 118 pieces or elements and the weight of the element is two times the atomic number of the element. Thus in truth, number one itself is composed of 100 percent plus 18!

That is to say, that the number one or even oneself is equal to 118 percent!

Also light or matter could enter into the vacuum of space or into the universe with only or as one element with its weight of two but that element would be able to convert into the other elements, even to the element 118 and its double weight of 236...

Element number one, for example, enters into the vacuum of space with its weight of two and it has 117 other possibilities of converting into the other 117 elements.

That is to say, element one is composed of 118 parts or pieces or the 118 percent and element number has 117 possibilities of converting into the other 117 elements according to the weight which element number one maintains. In the same manner, element number two with its weight of four has 116 possibilities of converting into the other 116 elements or until the element number 118 with its weight of 236...

When light or an element enters into the vacuum of space, light or the element enters as if it were a piece of magnet or as if it were a bar magnet.

In the vacuum of space the magnet or light or the element is super compressed not only until it takes the form of a sphere or round but also light or the element or the magnet is super compressed until it gets to a very high level of temperature.

And when the temperature gets to its highest level thus light or the magnet or the element super explodes causing the light or the magnet or the element to divide into two parts and the part with less weight, such as the negative part, takes position in the magnetic field and that magnetic field now is a negative field...

The superior or the positive or the heaviest part of light or of the magnet or of the element takes position in the center or in the nucleus. Thus, now we have the negative part of the magnet going around the positive part when before they were united and the center or the nucleus was neutral...

And even though the element lost weight because of the explosion or because of bursting, the element continues the one for two. That is, its weight continues of two although the element one now is 0.999 and its weight is double, 1.998...

Thus, now element number one was reduced to about 0.999 with its new weight of 1.998 but to element number one also remains a neutron or even more than one or a neutral part which can be converted or can be transformed into a positive part and that way not only adding the number of the element but also adding its weight even though it will have only one negative particle going around the center or the nucleus or the positive side...

But if one or light or the magnet or the element does not convert into the next number or into the next element thus it loses its energy and will only be a piece of dead matter in the vacuum of space and it will be removed one day by the black holes...

Thus, light or the element is the same as a negative particle, is the same as a neutral particle and is the same as a positive particle. In a way, one is equal to a negative portion plus a neutral portion plus a positive portion which totality is of 0.999 after entering the vacuum of space; but in the vacuum of space light or the element is positive even though the vacuum of space is neutral but reacts as if negative because of the vacuum.

And when the element increases its positive part by converting the neutral part into a positive part, the element cannot attract the negative part because of the vacuum of space because now the negative part becomes as if more or its weight increases because of the weight that it receives indirectly from the vacuum of space.

And if there were not positive attracting the negative thus the negative would expand through the vacuum of space and it would stop being negative and it would be dead matter...

Now, an element, in this case a star, which number is high as the same as its weight thus lasts or remains longer in the vacuum of space or in the universe. But the element or the star becomes heavier while the energy or matter to continue on lasts or it begins to transform from positive to neutral and once neutral, the element or the star practically becomes negative when its excess or super weight attracts the electrons toward the center and thus causing an implosion in where the element or the star becomes a nova or a new star but without light or without energy and that way causing an enormous hole in the vacuum of space when before the element or the star occupied the vacuum of space as an element or as a star...

In other words, matter or the star in the vacuum of space changes from positive to neutral and then from neutral to negative. Thus, + 0 -, in where the negative is a black hole or a super vacuum cleaner in the vacuum of space. This black hole practically eats or sucks all the matter around it to take all matter out from the vacuum of space or from the universe to make new space for new matter or for another beginning...

But as long as there are black holes in the vacuum of space or in the universe, thus the vacuum of space or the universe is negative and as long as the vacuum of space or the universe continues as or is negative thus it keeps being for something and not for neutral and from neutral or from zero to positive...

Thus, so that the universe becomes neutral or to zero and from neutral or from zero to positive thus all the black holes or super space vacuums must stop from functioning and once the black holes or the super space vacuums stop from functioning for lack of matter or for lack of energy thus the vacuum of space or the universe will return to neutral or to zero…

Now then, this new vacuum in space or in the universe is neutral and has zero energy but even so attracts new matter or new stars or attracts the new or the next beginning which is outside of the universe or outside the vacuum of space…

And once new matter or new stars enter into the vacuum of space or into the universe thus the vacuum or the universe will be positive or one or more…

Thus, the simple cycle or the model or the rhythm of (+ 0 -) continues until the end of all the times…

Chapter Three

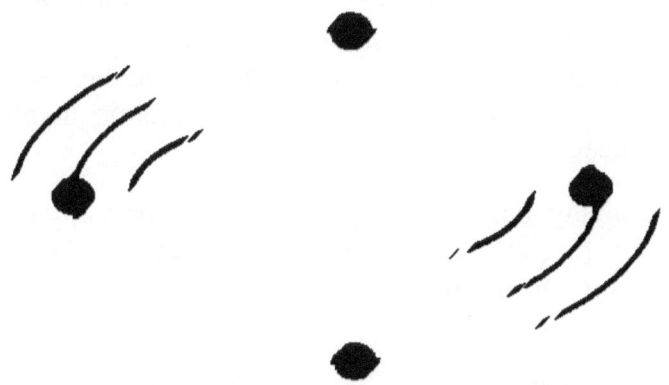

The Periodic Table of the Elements

Since all the solid pieces or since all the physical matter, the physical energy or the physical light from or in the physical universe or from or in physical creation came from one giant or from one infinite matter or from one infinite light or from one infinite energy cube, therefore, all the solid pieces or all the physical matter, energy or light in the vacuum of space or in inside the magnetic sphere or in the physical universe or in physical creation, must make or must resemble that giant matter, that energy or that light cube or at least make a solid part or a piece that would be very comparable to the infinite matter, to the infinite energy or to the infinite light cube.

All the physical matter, all the physical energy or all the physical light in the physical universe or in the vacuum of space, or in the magnetic sphere, is part of that giant or infinite magnetic cube which is the matter, the energy or the light cube; and putting all the physical matter, all the physical energy or all the physical light together in the physical universe or in the vacuum of space, that solid whole or that that solid part or that solid piece must be a cube or part of a cube.

All the physical matter, all the physical energy or all the physical light is actually broken into simple solids or broken into simple physical elements or pure matter, and not two elements or not two pieces of pure matter have the very same number of atoms or has the very same atomic number or atomic pieces.

It is like saying that no single puzzle actually or really has two identical or has two physical pieces that actually fit in the very same place or fit in the very same physical space at the very same time.

Turning to where the giant or the infinite space, the infinite coldness or the infinite darkness cube, and the giant or the infinite matter, the infinite energy or the infinite light cube physically unite, physically meet or physically interact, the very first small solid cube toward the infinite matter, the infinite energy or the infinite light cube is the very first element or is the purest or the simplest form of solid matter.

The very first element is simply called Hydrogen or element number one. H is the symbol for the Element Hydrogen.

The farther away from Hydrogen or the element number one, the heavier or the denser and the harder or the more solid the other elements.

That is to say, the further away is an element from the vacuum of space, the heavier or the denser and the harder or the more solid the element is.

The element or pure matter is also more pure since there is no physical interaction with the element or with pure matter outside the vacuum of space.

In short, the elements or the purest forms of solid or physical matter arranged or rearranged as small matter or element cubes and in accordance with their atomic numbers makes a matter, an energy or a light cube or makes half a matter, half an energy or half a light cube or makes a solid pyramid.

The solid pyramid is a three dimensional triangle or three parts of a solid whole or even a corner stone.

When a solid and a pure part or a solid and pure piece of the matter, the energy or the light cube enters the vacuum of space, the smaller solid part or the smaller physical piece actually carries the very same physical information as the whole cube or as the matter, the energy or the light cube.

When the physical piece or the physical part becomes even smaller, the smaller also carries the very same physical information as the whole or as the original infinite matter, the infinite energy or the infinite light cube.

A very excellent and a very unique physical example of the smaller solid part or the smaller solid piece carrying the very same physical information as the whole is the divisible atom or the element or even a very bar magnet.

If a bar magnet were to be broken or were to be divided to the very smallest possible physical piece, that very small possible physical piece would carry the very same physical information as the original or the whole bar magnet.

The divisible atom or the element or pure matter carries the very same physical information as the original or the parent matter, energy or light cube.

The very same physical process and the very same physical pattern or the very same physical information follows on forever even though some physical environmental issues are at hand or are involved.

For example, the physical change from matter, energy or light cube into matter, energy or light sphere as soon the matter, the energy or the light cube enters the vacuum or the compression of space.

And then from one giant matter, light or energy sphere into multiple and smaller matter, smaller energy or smaller light spheres through the Big Bang or through the Big Bangs!

There is also the fact that matter lost a lot of energy in the vacuum of space.

Energy lost is matter lost. Matter lost is space regained.

In reality, when a physical or a solid part or a solid piece of the matter, the energy or the light cube enters the magnetic sphere or enters the universe, the solid part or the solid piece is composed of four major parts: one part is matter and the other three parts are space or space-time.

So in reality there is a cube inside the magnetic sphere, in the vacuum of space or in the universe.

That is to say, the Universe's egg carries the very same physical or the very same solid information of its parent's cubes: one part physical matter, physical energy or one part physical light and three parts space or space-time.

In other words, the smaller matter, energy or light cube or the Universe's egg in the vacuum of space is composed of or made of one matter cube, one space cube, and two motion or two space-time cubes.

In short, there is one part physical matter, physical energy or physical light; and three parts space in the vacuum of space.

When the matter, the energy or the light cube in the vacuum of space turns or simply transforms into a matter, an energy or a light sphere, because of the vacuum or because of the enormous compression of space, the matter, the energy or the light sphere super explodes, thus losing its original size, original density or original mass.

The energy lost is really lost through neutrons or lost through the neutral part of matter.

The neutrons or the neutral part of matter in matter act as a protective shell just as an egg shell physically protects the egg.

When putting together the remaining physical or the remaining solid matter in the vacuum of space, that physical remainder should resemble a magnetic cube or half a magnetic cube or a magnetic pyramid.

Actually, a half magnetic cube, a magnetic corner stone or a magnetic pyramid will remain.

The simple reason that magnetic corner stone, half a magnetic cube or a magnetic pyramid remains is that the other half of matter or pure matter, energy or light was loss as physical energy in the vacuum of space.

Also, matter is made up of space-time.

That is, the other three parts of matter or the element is space and time, a part no seen but yet it is part of matter which is physical reality.

The elements or pure pieces of matter physically arranged as they actually or as they really are in the Periodic Table of the Elements does not really create a magnetic cube or half a magnetic cube or a magnetic pyramid or a magnetic corner stone.

The Periodic Table of the Elements actually or really looks rectangular.

That is to say, the Periodic Table of the Elements looks more like a bar magnet. See Figure Number Nine, the Periodic Table of the Elements.

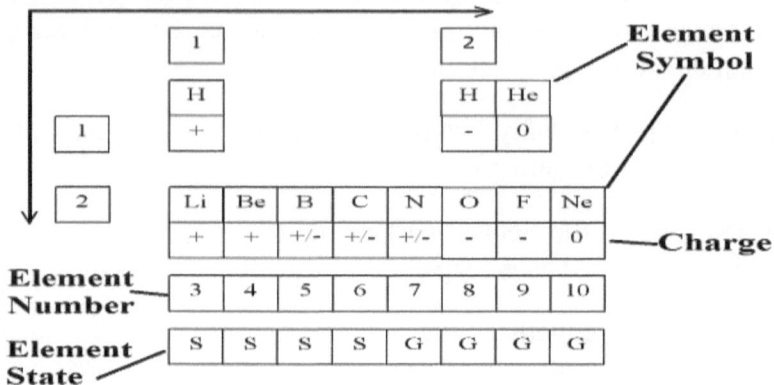

However, slightly rearranged, the Periodic Table of the Elements does make half a magnetic cube or a magnetic pyramid or simply a magnetic corner stone! See Figure Number Ten, Half a Cube.

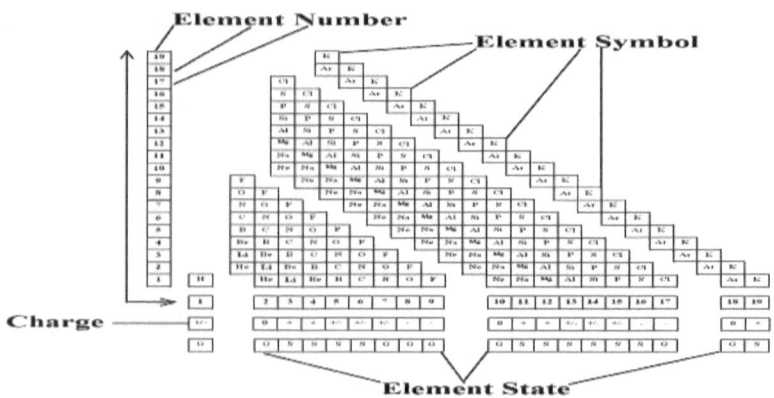

The Periodic Table of the Elements, which is simply a very simple diagram or a very simple blue print of a bar magnet, begins from left to right with a solid metal or with a solid, and ends with a gas.

The infinite or the original matter, energy or light cube begins from left to right with a gas and ends with a solid. The very simple reason that the infinite or the original matter, the energy or the light cube begins with a gas and ends with a solid is that the closer the infinite or the original matter, the energy or the light cube is to the vacuum or to the compression of space, the warmer or the hotter the temperature becomes due to the compression or to the vacuum of space.

The closer the infinite or the original matter, the energy or the light cube or the closer the simple element or pure matter is to the vacuum of space, the greater the compression or the greater the vacuum of space.

That is to say, the physical attraction or the physical magnetic force of the infinite matter, the infinite energy or the infinite light cube and the physical attraction of the infinite space, the infinite coldness or the infinite darkness cube is greater toward the intersection or greater in the center of both infinite cubes.

The physical force of attraction or the magnetic field is stronger at the intersection or stronger at the interaction of the two infinite cubes. However, with a single magnet or with a bar magnet, the physical force of attraction is actually or really stronger at the corners.

As one nears to or approaches the middle or the center of the bar magnet, the weaker the force of physical attraction or the weaker the magnetic field of the magnet.

The force of physical attraction becomes practically neutral or transitional, having both positive and negative, at the very middle or at the very center of the simple bar magnet.

The Periodic Table of the Elements, or table of atoms, also begins from left to right with a positive (+) charge, enters a transitional stage—having both positive and negative (-) charges—enters into a negative charge and ends with a neutral (0) charge.

The neutral charge is called in the Periodic Table of the Elements the rare or the inert gases. The neutral part of the Table of the Elements is actually where the Periodic Table of the Elements is divided or broken into simple sections. See Figure Number Nine, the Periodic Table of the Elements.

When using the infinite matter, the infinite energy or the infinite light cube or when using the magnetic pyramid or the magnetic corner as a table of elements, however, things are a little different; but this little or difference is what really makes possible the Periodic Table of the Elements.

The infinite matter, the infinite light or the infinite energy cube begins with a simple transition, enters a neutral charge, enters a positive charge, returns transitional, ends with a negative charge and begins again with a neutral or electrically neutral charge or ends with a break or with a division. See Figure Number Ten, Half a Cube.

It is the electrically neutral element that divides or that breaks the infinite matter, the infinite light or the infinite energy cube into seven different or into seven neutral parts or into seven neutral sections.

It is the seven different or the seven neutral parts or the seven neutral sections or the seven electrically neutral sections that make possible the seven periods or the seven divisions in the Periodic Table of the Elements.

It is also the electrically neutral element or the electrically neutral division that determines the number of electrons or the number of negatively charged pure matter particles in electron or in magnetic rings in the elements following the neutral element or following the electrically neutral matter.

The seven electrically neutral divisions or the seven electrically neutral parts is due to the magnetic field of the infinite matter, the infinite energy or the infinite light cube. The magnetic field or the magnetic sphere creates seven magnetic or seven electron rings around the infinite matter, the infinite energy or the infinite light cube.

In other words, the infinite matter, the infinite energy or the infinite light cube creates seven electron rings or creates seven magnetic rings around the infinite matter, the infinite light or the infinite energy cube.

The seven magnetic rings, as it can be seen, are neutral but when they enter the vacuum of space, the seven magnetic rings become electron rings. The transformation from neutral or from magnetic rings to electron or to negatively charged rings is due to the Bing Bangs of the divisible atom.

As soon as the divisible atom or the element enters the vacuum of space, the divisible atom or the element is super compressed. The super compression of the divisible atom or the element causes the divisible atom or the element to super explode, thus separating the electrons from the center or from the nucleus of the divisible atom or the element.

The electrons or the negatively charged pure matter particles take physical position in the field of the divisible atom or the element, transforming the magnetic field of the divisible atom into electron or into negatively charged rings.

Since the infinite matter, the infinite light or the infinite energy cube is also made of smaller pure matter cubes, the very first smaller pure matter cube is the first element or element number one. After the very first smaller pure matter cube, two smaller pure matter cubes follow: the second element.

There is one cube in addition to the very first cube. That is to say, first one cube, then two cubes, and then three cubes, and on until the last element on the Periodic Table of the Elements or on until the last element in the infinite matter, the infinite light or the infinite energy cube.

By the way, the very last element in the infinite matter, the infinite light or the infinite energy cube or in the Periodic Table of the Elements is the neutral element number 118.

All of the smaller pure matter cubes as simple elements or as pure matter or even as whole or as real numbers are diagonally arranged in the infinite matter, the infinite energy or the infinite light cube. See Figure Number Ten, Half a Cube.

In reality, the Periodic Table of the Elements and the infinite matter, the infinite light or the infinite energy cube or the magnetic pyramid or the magnetic corner stone are not too different.

The period or the section or the row of elements or pure matter begins soon after an inert gas: the neutral element or where the Periodic Table of the Elements or the infinite matter cube is divided.

There is a very simple physical transition or a very simple physical change from gas to liquid and then to solid. However, the Periodic Table of the Elements does not show that simple physical transition or that simple physical change.

Instead, the Periodic Table of the Elements almost shows the exact physical opposite; but it must be well understood that the Periodic Table of the Elements is simply affected by Earthly temperatures and Earthly pressures.

What this simply means is that the infinite matter, the infinite light or the infinite energy cube begins with a gas, enters a liquid stage and ends with a solid; begins again with a gas, re-enters a liquid stage and ends with a solid.

In reality, the infinite matter, the infinite light or the infinite energy cube is solid as metal or as solid steel, but the closer the infinite or the original matter, the energy or the light cube is to the infinite space cube or to the vacuum or to the compression of space or the closer to the magnetic sphere, the softer or the less solid and the lighter or the less dense the infinite or the original matter, the light or the energy cube actually becomes.

The further away from the vacuum or the further away from the compression of space, the heavier, the more solid or the denser the infinite or the original matter, the light or the energy cube becomes.

The closer to the belly or to the middle of the two magnetic cubes, the softer is the matter and the weaker the magnetic attraction.

Matter, energy or light really is composed or made up of three main physical or transitional parts. The three main physical or transitional parts are positive, neutral and negative or simply put, "+0-."

That is, M=+0-. In other words, matter, which is reality, is plus zero minus.

Zero charged or the neutral part of matter unites and divides as in the Periodic Table of the Elements. So in reality all of matter is transitional or really turns from positive to neutral to negative. Or matter changes from one element to another.

That is to say, the simple plus zero minus allows matter or an element to change into another element or into another state. That is, from solid to liquid to gas, and vice versa.

As it can be seen, the Periodic Table of the Elements is not only a very simple diagram or a very simple blue print of a very simple bar magnet or a very simple diagram or a very simple blue print of an electric motor or a generator, but the Periodic Table of the Elements is also the simple blue print of all Existence.

When putting or reconnecting pure matter back together using its atomic number or using the simple elements as in the simple Periodic Table of the Element, after the Big Bangs!, half a magnetic cube, actually half a matter, light or energy sphere, or a magnetic corner stone or a magnetic pyramid is left.

The other physical part of pure matter, pure energy or pure light that is missing was lost as pure energy in the form of neutrons in the Big Bangs!

For every one divisible atom or for every one particle of pure matter, pure energy or pure light that physically entered the vacuum of space or that physically entered the magnetic sphere or that physically entered the universe and entered the other universes, one neutron was lost as energy.

For example, Element or pure matter particle number one or atom number one lost one neutron as energy; but the very last Element or the Element number 118 lost 118 neutrons as pure energy.

By the way, there are really only 118 Elements or only 118 pieces of pure matter, pure energy or pure light or only 118 divisible atoms in all of existence or in the universe or in all the universes or in all the dimensions of time.

Also, the atomic weight of the divisible atom or the actual weight of an element is twice the atomic number or twice the weight of the element.

For example, for element number one or pure matter one, the weight is two; and for element number two, the weight is four; for element number three, the weight is six, and so on until the very last element, Element 118.

The Elements and Their Weight

E/W	E/W	E/W	E/W	E/W	E/W
1=2	21=42	41=82	61=122	81=162	101=202
2=4	22=44	42=84	62=124	82=164	102=204
3=6	23=46	43=86	63=126	83=166	103=206
4=8	24=48	44=88	64=128	84=168	104=208
5=10	25=50	45=90	65=130	85=170	105=210
6=12	26=52	46=92	66=132	86=172	106=212
7=14	27=54	47=94	67=134	87=174	107=214
8=16	28=56	48=96	68=136	88=176	108=216
9=18	29=58	49=98	69=138	89=178	109=218
10=20	30=60	50=100	70=140	90=180	110=220
11=22	31=62	51=102	71=142	91=182	111=222
12=24	32=64	52=104	72=144	92=184	112=224
13=26	33=66	53=106	73=146	93=186	113=226
14=28	34=68	54=108	74=148	94=188	114=228
15=30	35=70	55=110	75=150	95=190	115=230
16=32	36=72	56=112	76=152	96=192	116=232
17=34	37=74	57=114	77=154	97=194	117=234
18=36	38=76	58=116	78=156	98=196	118=236
19=38	39=78	59=118	79=158	99=198	
20=40	40=80	60=120	80=160	100=200	

E210 + 610 + 1,010 + 1,410 + 1,810 +1,971=7,021=10=1

W420 +1,220 + 2,020 + 2,820 + 3,620 +3,942=14,042=11=2

E118=1; That is, all of the elements equal one unit or one unit equals 118 parts.

In other words, one is equal to 118 parts...

W236=2; that is, the weight of an element will ways be twice the element number even if that number transforms into another element...

Chapter Four

The Divisible Atom

What is, therefore, the divisible atom?

The divisible atom is a microscopic or is a physical miniature of the infinite matter, the infinite light or the infinite energy cube and the infinite space, the infinite darkness or the infinite coldness cube.

The divisible atom is a microscopic or is a very physical miniature of existence or of the universe or of physical creation.

The divisible atom is a microscopic magnet inside a magnetic sphere or in a magnetic field composed of seven electron or seven magnetic rings.

The atom as microscopic matter is also composed or made up of three main physical parts.

The three main physical parts are protons, neutrons and electrons or simply put, "+0-."

That is, A=+0-. In other words, the atom is plus zero minus.

Atoms are also transitional or turn from positive to neutral to negative.

Contrary to its name, the atom is physically divisible or is physically broken into three main physical parts.

And those three main physical parts can also be divided or can also be broken into many smaller physical and identical parts as the first three main physical parts.

The divisible atom is also made of protons (+), neutrons (0) and electrons (-). The seven electron or the seven magnetic rings of the divisible atom are created by the atom's magnetic field or shield or the atom's magnetic sphere.

Actually, it is an atomic magnetic shield or an invisible magnetic or electron shell. See Figure Number Eleven, The First Three Rings of the Atom.

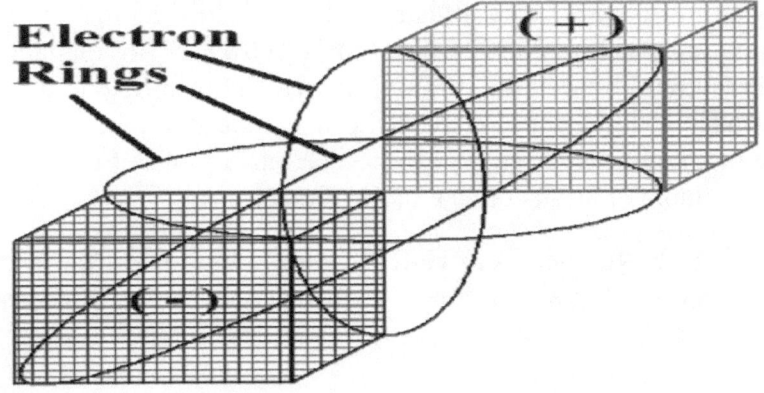

And see Figure Number Twelve, The Seven Electron Rings of the Atom.

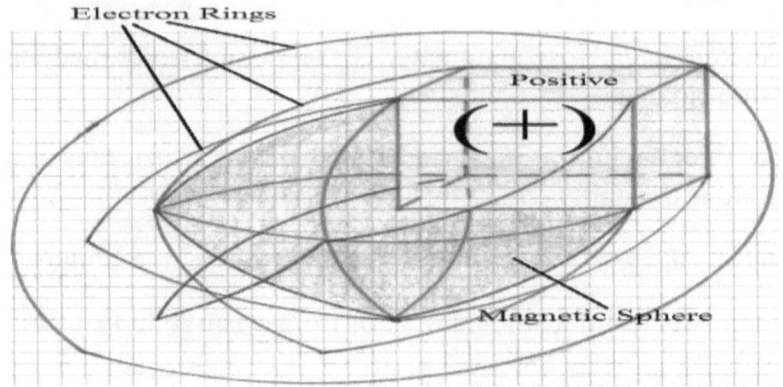

When the divisible atom or when pure matter or when the element physically enters the vacuum of empty space, the divisible atom is physically divided or physically broken or is physically separated from the divisible atom's center or from the divisible atom's neutrally charged nuclei or neutrally charged center, and the electrons occupy a very specific physical part or a very specific physical location or occupy a very specific physical point or occupy a very specific physical position of the magnetic field or the now electron ring.

What this actually or simply means is that the divisible atom went through a microscopic or a miniature physical expansion or the divisible atom simply went through a microscopic or a miniature big bang!

To show or to simply illustrate the divisible atoms or the miniature pure matter particles in electron or in magnetic rings, the infinite or the original matter, the infinite energy or the infinite light cube or even the magnetic pyramid or even the magnetic corner stone can be used as a table of elements or as a table of atoms or as a table of electrons.

See Figure Number Ten, Half a Cube below or on the next page; and see Figure Number Thirteen, Table of Electrons/ the first three rings.

The new table of elements, the new table of atoms or the new table of electrons is easily read diagonally from left to right and from bottom to the top.

However, the new table of atoms or the new table of electrons can also be easily read from top to bottom, but only when the table is flipped or turned upside down. The very first cube or the very first square in the new table is the first atomic number or the first element or element number one, which is the element Hydrogen. H is the symbol for the element Hydrogen.

Hydrogen has one proton or has one positively charged atomic or has one microscopic particle or has one atomic matter in the center or in the nucleus or nuclei of the divisible atom. The element Hydrogen also has one neutron or has one electrically neutral atomic matter particle in the nucleus. The element Hydrogen also has one electron or has one negatively charged atomic or has one microscopic particle or has one atomic matter on the first electron or the first magnetic ring.

However, before Hydrogen enters the vacuum of space, Hydrogen has two neutrons or two electrically neutral particles of pure matter.

When the element Hydrogen enters the vacuum of space, the element Hydrogen really losses one neutron as pure energy!

In fact, all matter or all elements or all atoms lost neutrons as pure energy in the vacuum of space. Pure matter, the elements or the atoms have twice the number of neutrons before pure matter or the divisible atom enters the vacuum of space.

For every one proton or for every one electron that enters the vacuum of space, one neutron is lost as pure energy in the form of heat. One neutron for each atom was lost. Pure matter, the elements or the atoms lost neutrons according to the atomic number.

For every proton or for every electron, one neutron is lost as pure energy in the vacuum of space. The loss of neutrons as pure energy in the vacuum of space is 1, 2, 3, and 4, and so on until the very last atom or the very last element in the Periodic Table of the Elements.

The very last atom or the very last element is atom or is element number 118. So, element or pure matter number 118 in the vacuum of space lost 118 neutrons as pure energy.

Helium, the second element on the Periodic Table of the Elements or the second piece of pure matter in the new table, has two cubes or two physical parts of atomic matter. The symbol for Helium is He.

What this means is that Helium has two atoms or has two atomic particles of pure matter. That is to say, the two cubes in Helium means that Helium has two protons or has two positively charged pieces of pure matter. Helium has also two neutrons or has two electrically neutral particles of pure matter in the center or in the nuclei.

For every one proton there is a neutron and there is also an electron. It is a simple diagram of a bar magnet, in where a proton is on one side, the neutron is on the middle, and the electron is on the other side. And because of the Big Bang the three main parts or the three main solid pieces are separated from the center or separated from the nucleus. The proton and the remaining neutron stay at the center or at the nucleus, but the electron takes position in the magnetic field or in the so-called electron or magnetic ring.

Actually, the magnetic field or the magnetic ring is really a neutral or is an electrically neutral ring or field. When the electrons are separated from the center or from the nucleus of the divisible atom, the electron occupies the neutral ring making or turning the neutral ring into an electron ring or electron field or shield. The electron transforms or charges the magnetic ring or shield into an electron ring or into a negatively charged ring,

To simply find the exact number of electrons or the exact number of negatively charged pure matter particles orbiting or going around the nucleus of an atom, one must count by two for the very first electron or for the very first magnetic ring and whatever number of electrons or whatever number of negatively charged pure matter particles remains goes to the second electron or to the second magnetic ring.

Helium, for example, has two electrons or has two negatively charged pure matter particles on the very first magnetic or the very first electron ring or the very first magnetic or electron orbit.

In reality, the Element Helium is an element made up or composed of two magnets united at the corners as the infinite matter cube and the infinite space cube are united. And being made up or being composed of two magnets, the Element Helium has two protons, two neutrons and two electrons.

Turning to the third element or the three diagonal smaller cubes or the three smaller squares or the three whole or real numbers, count two electrons for the first electron or the first magnetic ring and count one electron for the second electron or the second magnetic ring.

And finally, turning to the fourth element or to the fourth divisible atom, count two small cubes or count two small squares for the first electron ring, and there is a remainder of two smaller cubes or two smaller squares for the second electron ring. The fourth element is really four magnets united at a corner or united in links.

The very same example or the very same and simple physical pattern as above follows on until the next neutral element or the next neutral atom, which is a gas. Pure matter as gas or an element as gas is not electrically charged.

Also, the table of atoms or the table of electrons or the infinite or the original matter cube is divided at the neutral or divided at the electrically neutral element. See Figure Number Ten, Half a Cube; and see Figure Number Thirteen, Table of Electrons/the first three rings.

The next neutral element, which is Neon or element number ten, follows the very same and simple example or simple physical pattern of the first elements before Neon: count two small cubes or count two small squares for the first electron ring or the first magnetic ring, and count eight small cubes or count eight small squares for the second electron ring or for the second magnetic ring. The elements following Neon or following element or atom number ten follow the physical pattern of Helium, two electrons, and Neon, eight electrons.

The pattern for Neon is two small cubes or two small squares for the first electron ring, eight small cubes or eight small squares for the second electron ring and whatever number of smaller cubes or whatever number of smaller squares remains goes to the third electron ring or goes to the third magnetic ring.

Element number eleven is counted two small cubes for the first electron ring, counted eight small cubes for the second electron ring and counted one small cube for the third electron ring or the third magnetic ring. Element or atom number twelve is counted two small cubes or counted two small squares for the first electron ring, counted eight small cubes for the second electron ring and counted two small cubes for the third electron ring or the third magnetic ring. See Figure Number Thirteen, Table of Electrons.

The very same physical example or the very same and simple physical pattern as above follows on until the next neutral element or the next division of the table, and the very same and simple physical procedure as Helium and Neon must also follow.

That is to simply say, the next neutral element determines the number of electrons in electron rings or in magnetic rings. The next physical break or the next physical division of the new table, the Periodic Table of the Elements or the infinite or the original matter cube determines the exact number of electrons that are going to the electron rings or to the magnetic rings. It is already pre-determined even before the original matter cube or the original atomic cube enters the vacuum of space.

As it can also be simply seen, the divisible atom is not only a very simple microscopic diagram or a very simple microscopic blue print of a simple bar magnet or a simple microscopic diagram of an electric motor or generator, but also the divisible atom is a very simple microscopic diagram of the physical universe and of physical Existence or of physical reality or of physical creation.

The divisible and the physical atom is actually a simple magnet inside another simple magnet, but instead of the two simple magnets physically attracting or physically uniting as two opposite charged magnets, the two atomic or the two microscopic magnets are kept physically apart by the magnetic sphere or by the magnetic field or shield or simply by a magnetic shell.

In a way, the magnetic sphere or the magnetic field of the divisible atom acts as a simple protective shield or acts as a simple protective sphere or act as a simple protective shell.

The simple magnetic sphere or the simple magnetic shell, therefore, keeps the divisible atom from physically collapsing or from further physical expansion or from further physical separation from the nucleus or from the center of the divisible atom. That is the simple reason that the element or that pure matter does not become physically much smaller or microscopic as the divisible atom.

In other words, the simple magnetic field or the simple magnetic sphere or the simple electron ring of the divisible atom or the element keeps matter from further physical expanding or from physical collapsing or from physically becoming microscopic or from even physically imploding.

However, this does not mean that the magnetic sphere or that the electron ring cannot be acted upon from outside. In contrast, interacting with the divisible and the physical atom from the outside creates mixtures of new elements or two or more elements united, thus making other elements or what is simply called in Modern Chemistry isotopes, compounds or mixtures.

This, however, does not mean that there will ever be more than 118 elements.

The Small which makes the Great

Existence is infinite or without size. Existence runs or she expands toward all sides at the very same time and at the very same time existence also compresses toward all sides, that way also adding to her very self an infinite weight and an infinite size.

That is to say, existence adds to herself, she multiplies herself, she divides herself and existence also subtracts her very self or from her very self to be able to become forever for more and even for new.

Existence forever existed and existence forever will exist. Existence has no time even though time existence is. Existence is also composed of three principal parts which all add to one and that one is the grandiose part which makes the difference in all of existence.

Existence consists of matter, that which is physical. Existence also consists of space or of vacuum, that which is lack of something or of matter. And existence also consists of movement or of that which many call space-time, which is a movement as if in waves in space.

Thus in truth, existence is a simple magnet!

In other words, existence really is composed of +0-; in where (+) is equal to positive; in where (0) is equal to neutral; and in where (-) is equal to negative or the lack of. In the scale of colors would be white, grey and black; in where white is light or matter; grey is neutral or space-time; and black is darkness or lack.

Now then, just as existence truly is composed of three parts and practically of parts opposite to the other parts, +0-, thus one also truly is composed of those very same opposite parts but in knowledge because one is knowledge.

And one can truly be positive knowledge or be neutral knowledge or even be negative knowledge. But the greater the positive knowledge one has, less the neutral knowledge and less the negative knowledge or less the lack of negative knowledge.

In other words, the greater the knowledge of reality that one has, the greater the reality of one and less is the reality not known or less is emptiness or darkness.

Thus in truth, the more the knowledge of one, the more the abundance or the more is the life of one or more life makes sense to one.

Interestingly, that existence functions the same as one but existence knows it not but existence will never stop from being or will never stop from existing for all the eternity of eternity because she truly makes herself for much more!

That is in truth, even though all of existence is one, because of one existence truly is and for all eternity infinite!

And every time that existence is increased by one or is increased by more than one, the point of neutrality or the neutral is less as the same as the negative or the unknown point is less.

But making the things less does not make existence greater, but making the things or existence greater makes existence greater!

And just as existence makes the things greater for existence to be greater herself, oneself can be greater by recognizing that life is more or that in life herself there is more and that there is with all peace, the neutral, and that here is with all knowledge, one or more than one; and also that there is with all gladness and with all joy and that also there is with all abundance of life, life renewed...

Thus, through knowledge one renews for much more and because of one renewing for much more, one truly can continue for all of eternity as the very eternity...

Creation and the Number

Creation truly is about giving name or naming to start or to begin and giving rename or renaming to continue without beginning again even though to give rename or renaming is to become as if forever new and as if never there were a start or a beginning and neither end...

All of existence truly is composed of numbers or be it positive numbers or be it neutral numbers or be it negative numbers or be it a combination of all the numbers at the very same time, but there will always be a number that is greater than the other negative numbers until that positive number is converted into a neutral number and after that positive number is converted into a neutral number thus also it will be converted with time into a negative number.

In other words, existence really is equal to (+ 0 -), but existence does not return to negative but rather a part of her and that part of her is the universe or matter or light, which returns to or truly is converted from positive to neutral or to zero and much later it is converted to negative or into nothing. The number zero is not negative or nothing. The number zero only is a neutral number which could be converted into a positive or negative number.

Existence as matter or as light is truly composed of three parts which add to 118 percent, 118 times 3 because they truly are 118 positive parts, 118 neutral parts and 118 negative parts or (+ 0 -).

Now then, the neutral part or part zero is the universe and it is where it is added, it is subtracted, it is multiplied and it is divided at the very same time. The positive part is the physical part of existence or the part from where comes out matter or light or the elements which truly are pure matter, but this part of existence appear to be minor than the other parts of existence.

The negative part of existence is the lack or is the vacuum of space or darkness which is composed of three parts, one is space or emptiness and the other two are space-time, which truly are created by the interaction of the magnetic field between the positive and the negative.

When matter from the physical side of existence enters into the vacuum of space, matter even though a single piece, matter enters with its 118 parts assimilating the 118 elements.

That is, if only one element enters into the vacuum of space that element could assimilate or even can come to be converted into the other 117 elements...

But in reality matter enters from the physical side of existence into the vacuum of space in 118 pieces into 118 parts of the vacuum of space or into 118 dimensions at the very same time.

That is to say, into the vacuum of space there enter 118 elements and every element has 118 parts and at the very same time there enter 118 elements into 118 dimensions. But that does not remain like that, because with every element the 118 parts are multiplied.

In other words, element number one has 118 parts but element number two has 236 parts and element number three has 354 parts and this pattern continues on until the last element, which is element number118.

Also, the weight of the element is twice its number. That is to say, element number one has a weight of two and element number two has a weight of four and element number three has a weight of six. This pattern also continues on until element number 118.

Interestingly, that all the parts of an element add to the number of that element!

Thus, element number one has 118 parts and if we added those 118 parts thus we would get one. That is, if we added 1+1+8 thus they would give us 10 and if we added 1+0 thus we would get 1. In the case of element number two, if we added 2+3+6 thus we would get 11 and if we add 1+1 we would get 2. This pattern continues on until element number 118. Element number 118 has 118 parts and every part also has 118 parts thus giving a sum of 13,924, also adding to one.

Now then, when element number one enters into the vacuum of space element number one enters with one proton, with one neutron and with one electron, (1+0-1). And also element number one enters with a weight of two.

But once in the vacuum of space, space comprises the element and the element is comprised causing the element to super heat and thus also causing the element to burst or explode.

In that burst or explosion the element loses a neutron or a third part and no longer the weight is of two but is less, such as 1.67.

Also the electron was separated and now is spinning around the proton or the nucleus of the element, (+) -.

In the case in where an element has a high number thus that element will neutrons in its nucleus, such as (+0+) -, -. Or (+0+0+) -, -, -. This new transformation of the element, in where the electron is separated from the nucleus makes it possible for the element to be united to other elements and that way converting into a mixture of elements or isotopes…

Furthermore, before an element or matter enters into the vacuum of space its three main parts have the same size or weight, ((+) (0) (-)).

But once that element enters into the vacuum of space thus that elements is divided into two parts, the nucleus in where new is the proton (+) and the neutron (0), and the outside part in where now is found the electron (-) going around the nucleus, ((+) (0)) (-).

And while the nucleus keeps its size or weight, the electron loses its size or weight because of the interaction or friction which it has with the nucleus or with the center of the element. The interaction or friction which the electron has with the nucleus or with the center of the element also causes the electron to last less or lasts less time in the vacuum of space.

Once the electron is fused or is exhausted, the element or the nucleus is converted into a neutral element or without energy even though the nucleus is still positive or with protons and neutrons.

But the outside part or the electric field of the element is now a neutral field or the electrons have been converted into neutrons because of lack energy.

In other words, the seven electron rings of the element or the atom now are neutral.

And just as the element functions thus that way also functions the number and existence herself, but the number or the symbol of the number is only an illustration of the numbers but truly does not show how is the number in existence or outside the vacuum of space in where there is no friction or movement even though there is a magnetic field.

Thus in truth, the number one or 1 outside the vacuum of space is represented by a cube. Now, the cube or the number or the element one is composed of three main parts and they are the positive part, the neutral part and the negative part.

But those parts also are composed of 118 other parts. That is to say, that the positive part also is composed of 118 parts and the neutral part is also composed of 118 parts as the same as the negative part which also is composed of 118 parts.

And when the cube or the number enters into the vacuum of space the cube or the number is compressed into a sphere or into a globe but still with its 118 parts.

Thus, the number one is composed of not only 100 percent but also of another 18 parts or of another 18 percent. And if we added the parts thus we would have 1.

Now then, the number one or the symbol 1 as also is all of creation is a continuation because the start or the beginning is zero or a point or a neutral or an empty space.

But the number one or the symbol 1 also represents all of existence, the physical part, the neutral part and also the negative part.

Also the number one has the ability of converting itself into its 117 other parts also with their other 118 parts.

In other words, the number or element one also has the ability of being infinite because also it could renew into a greater number such as the number two.

And that makes it possible the other 117 parts which add to 9 and the number 9 is a symbol of renovation. That first renovation extends the time of the number or of element one.

Thus, if we added the 117 parts which remain to one plus its other two parts, the neutral with its 118 parts, and the negative with its 118 parts, the sum would be of 353.

And if we added 353 thus it would give us 11 or eleven and if we added 11 we would get two, the possibility or the ability of the number or element one if it is renewed.

And once that the number or that element number one has renewed as two, thus the number or element one has become or will continue as double or for much more as two and as double the abundance.

And the very same step or process is with the number or with element number two. If we added all the parts that the number or the element two has, which are the double of one, thus it would give us four or 4.

That is to say, if we added all the parts of the number or element two thus we would get the double.

Chapter Five

Summary

Existence simply is. Existence always was. And Existence will always be.

In short, all of Existence, physical or not, always existed forever and ever. Nothing or lack or nothingness is really also something. Nothing, lack or nothingness is also a major part of Existence.

Actually, Existence is really made-up of three parts which really are not or three parts which simply do not exist as physical Existence really exists. Nothingness, lack or simply non-existence also exists but non-existence does not really make possible physical Existence.

Non-existence or lack does not really or does not actually happen by itself and non-existence or simply non-reality or lack does not make possible physical Existence or does not make possible actual or physical reality even though non-existence or lack or nothingness is really or actually a reality.

That is to simply say or the simple reality is, non-existence or nothing or nothingness or even lack does not really happen by itself. Non-existence, nothing or nothingness or non-reality or even lack really requires or really needs physical Existence or really needs or requires physical or actual reality.

In short, non-existence or nothing or nothingness or non-reality or even lack, therefore, really or actually requires or really or actually needs physical something or something solid.

That is, non-reality requires physical reality. And nothing, non-existence or lack is something in itself and, therefore, exists. Once again, nothingness or non-existence or even lack is a reality, although not physical or actual reality.

However, even though non-reality or nothing or non-existence or even lack simply exists, nothing, non-reality or non-existence or even lack really cannot exist by itself. Nothing or non-existence or lack needs or really requires something or nothing needs or really requires all of Existence to exist. Physical Existence or physical or actual reality is really the simple product of Existence and non-existence. Physical Existence or physical or actual reality, therefore, completes non-existence.

That is to simply say, physical something or physical reality really or actually completes or really or actually makes perfect nothingness or nothing or makes perfect non-existence or makes perfect even lack. Nothing, lack or non-existence is not complete or not perfect. Only physical reality or the naked truth can really or can actually complete or can really make perfect non-reality or can cancel lack by filling or by completing lack.

Physical Existence or physical something or physical reality or actual reality really makes possible non-existence or makes possible non-reality or makes possible lack. Contrary to common or to popular believe, physical something or physical reality or physical existence did not come out of that that was not. Physical existence already was or was always present at all times.

And of what already was present all the times, more came out or more came to be. Nothing or lack really or actually attracts that which already is and not what is not.

In other words, the simple shadow or the simple shade is not because of lack of light. Physical Existence attracts nothing. Physical existence simply or actually or really attracts nothing or actually or really attracts non-existence. Physical existence does not attract physical existence because physical existence is already complete or already full or physical existence simply is physical.

Only opposites really or actually attract. Non-existence or nothing or even lack really does not attract what is not. Non-existence or nothing or even lack really attracts what is.

Physical something or physical reality does not attract itself because something is already physical reality, but physical something or physical reality actually or really attracts that which is not. Non-existence or non-reality attracts physical existence or physical reality since non-existence or non-reality is not physically complete or is not physically perfect.

Even though nothing or non-existence or lack is a reality that exists, nothing or non-existence or lack is a reality that lacks physical reality. And lacking physical reality, nothing or non-existence or lack is, therefore, not complete even though nothing or non-existence or lack is greater in size than physical reality or physical existence.

That is to simply say, no real or true question no matter how small or how simple is truly or really answered unless first asked and then a truthful or a real answer is really expected.

The very simple question or point of view is really or actually recognizing or really remembering or really recollecting that there is but more!

Something or reality really goes into nothing or into lack, and not the other way around. Nothing or lack can be filled if it already is filled. Occupied space is not space. Occupied space is really the physical or the actual universe. And if nothing or if lack already is filled, then it is physical something or actual reality and not lack or nothing.

That is to simply say, if it is filled, then it is not empty and not lacking. And such is the simple rationality of all physical or actual Existence or actual reality. And so the closer to the rational, no more irrational to the conscious living being or to conscious thought.

Physical existence or physical something or even physical reality speaks through physical creation, but physical existence or physical something or even physical reality speaks with a double meaning which too often adds to physical existence or too often adds to physical something or too often even adds to physical reality through too much contradiction. So even actual or physical reality is a point of view or a way of thinking!

And so the more rational to some thinking being the more irrational to some thinking being. Nothing seems like it is because nothing is unseen. Nothing is a reality not really seen. Nothing is the lack of physical reality.

For example, to go up is also to go down; to hang upside down is also to hang right side up; to pull is also to push; to leave is also to arrive; to come is also to be already present; to simply multiply is also to simply divide; to add is also to subtract.

As one moves forward one step, one has also one less step to move back or forward; and also to point one way is also to point to many ways or directions; to close one's eyes simply is not to stop seeing.

But because no rational understanding or no rational knowledge is really found in the irrational by itself, which is really a major part of the rational, the other percent of that that already is here and at once cannot be really or actually seen. Remember the animal mind and the tree?

Although the simple tree is entirely present and complete with fruit and all, the animal mind cannot see the simple tree because of the simple point of view of the animal mind.

Knowledge simply but is or exists. Knowledge is also physical or a physical reality. Knowledge as physical existence always existed and physical knowledge will always exist, even without conscious living beings.

Learning is, therefore, really simple recollection or actually or really remembering and remembering is really completing. The greater the point of view of the conscious living being is, the greater the learning or the greater the simple recollection or the greater the remembering. The greater the learning or the greater the actual or simple recollection or remembering, the greater or the more complete Existence or actual reality will thus be.

The universe and Time are simple products of non-existence and Existence. That is to say, the universe and Time are products of two simple opposites: space and matter, negative and positive, non-existence and existence, vacuum and expansion, black and white, that which physically already is and that which physically is not.

Where those two opposites physically meet or physically intersect or even physically divide, there is the physical universe and physical motion or Time. Time, which is physical motion or is physical expansion, is simply created by the simple magnetic fields or simply created by the physical magnetic attraction of two simple opposites: space and matter.

The space cube and the matter cube or simply all of Existence is a simple electric motor, really a magnetic motor or generator. The electric or the magnetic motor is made possible by a very simple magnetic field or a magnetic shield.

In other words, the simple magnetic field makes possible the physical motion in the motor.

In the same way, the magnetic field of the two magnetic cubes makes possible motion and Time or expansion.

Time, like the physical universe and physical Existence, always was and will always be. Time, which is simply physical motion or simply physical expansion, had no physical beginning and will have no physical end.

The physical motion or the physical expansion in Time is created by the simple magnetic field of the two magnetic cubes.

Time, therefore, is a simple point of view. Any physical beginning is simply a physical continuation of a physical existence that already was.

To go in, for example, is a continuation from something that already was. There was not really a first or an actual physical beginning, but a physical continuation because physical existence already was.

The only true or the only real or the only actual beginning is, therefore, that of conscious thought.

Conscious thought, therefore, really completes physical reality or completes physical or actual existence.

That is, conscious thought really or actually makes or really or actually completes physical existence or actual reality.

Conscious thought really or actually makes possible the reality not seen!

There were 118 physical beginnings that all began at once with 118 endings that all ended at once.

That is, if something began and ended at the very same moment, can it be called beginning?

Can it be called end?

The beginning and the end are but the very same thing. They are both really a continuation, and they are both but a very simple point of view.

For example, when leaving, one is also arriving at the very same time even though the actual arrival or the physical point of view is much later or the actual arrival or the physical point of view really will be completed or perfected much later.

In other words, the actual arrival is a very simple continuation of leaving. Even aging or growing old is a simple physical or actual continuation or expansion or even a point of view.

Once again, when turning left, one is also turning right, turning south and turning north at the very same time and at the very same location since physical motion is simply a continuation.

When falling or when moving down, one is also going up to the ground!

North, South, West, East are really different names to the very same thing or the same place and that place or thing is also present at the very same time and at the very same place or space; and those names are also different names to a very simple and the very same point of view.

There is no such thing as a true or as a complete or a perfect north, a true or a complete or a perfect south, a true or a complete or a perfect west or a true or a complete or a perfect east, no matter how straight the magnetic needle points.

All of those points of location or space, once again, are really present in the very same physical location at the very same time.

That is to simply say, everything exists at once, but not everything is really seeing at once simply because the conscious living being limits his conscious thinking, simply limits his way of thinking or simply limits his point of view. See Drawing Number 14, The Point of View or the Origin of the Opposites.

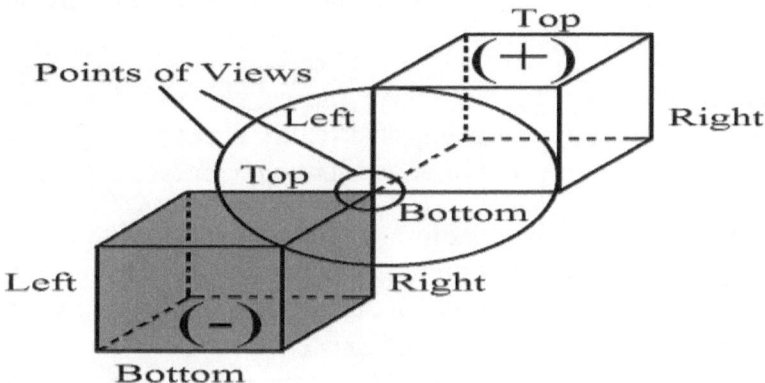

According to the naked knowledge or to the understanding or even the simple recollection or acknowledgement that the conscious living being really or actually has of Existence, the universe or physical creation, Existence, the universe or physical creation will thus be.

In other words, the greater the point of view or the greater the recollection, the greater Existence or the greater the physical creation will simply be.

So in reality, Existence equals plus zero minus. Or simply put, "R = +0-." Reality or physical creation really depends on the point of view of the conscious living being.

Once again, yesterday, today and tomorrow are also points of view. They are really the very same thing and they are also present at the very same time.

That is to simply say, tomorrow is already spent and tomorrow is only a point of view. Yesterday is a future to another yesterday.

Actually, yesterday was many futures. And yesterday is still here today. Today is a future to many a past. Once again, the simple reality here is that today is multiple futures to many a past. Tomorrow, though a future of many a yesterday and a future of today which is another future, tomorrow is also a past and tomorrow also is already spent. Yesterday, today and tomorrow are all really present at the very same time here and now.

Thus, reality depends on where the conscious living being stands. That is, actual or physical reality depends on point of view. And what is point of view if not but conscious thought?

Time or beginning is, therefore, a simple point of view. Time or beginning or point of view is a simple continuation of that which already is, of that which already was and of that which will always be, and that is all of Existence.

All of Existence really is present all at once and in the very same space or location. The most interesting thing is that when the conscious living being gets to that point or that location or space, where everything or all of Existence meets at once, is that nothing will really be seeing and that nothing, not even anything, will really make any sense. In other words, all of Existence in itself or by itself really makes but no sense.

A physical part or a physical piece has to really or actually disengage or has to really or actually depart so that the physical greater or the physical larger part or physical infinity can be seeing through the physical smaller or through the physical lesser part; and also that the physical smaller part or the physical finite can be seeing through the physical greater or can be seeing through physical infinity.

To simply count or to simply add, therefore, the conscious living being has to simply take apart or has simply to divide. And ironically, really multiply and subtract all at once or at the very same time.

The only thing that the conscious living being should really know or should really understand or should really realize or simply or actually recollect about time is that the conscious living being is limited to time or limited to physical expansion or simply limited to point of view only because the conscious living being limits himself through conscious thought.

And that the conscious living being must, through conscious thought or through conscious effort, change or improve his point of view or change or improve his way of conscious thinking and thus be able to complete or to make perfect physical existence or complete or make perfect physical creation or the conscious living being simply dies, therefore, allowing for a second physical beginning or a second physical creation of the universe and universes.

However, even though a second physical beginning or a second physical creation will be twice the size of the first physical beginning or twice the size of the first physical creation with even bigger Big Bangs, that is no reward.

A second physical beginning or a second physical creation is no second chance. A second physical beginning or a second physical creation is really or actually the failure to complete or the failure to perfect the first physical beginning or the first physical creation (through remembering) or through acknowledgement.

By the way, those conscious living beings living and present here in this universe or the other present universes will not be present or will not come on the next physical beginning or to the next physical creation. Every physical beginning or every physical creation is really a one-time deal and only one opportunity to complete or to make perfect that physical beginning or that physical creation and not any other physical beginning or not any other physical creation.

And as long as there is conscious thought in the present universe and present universes, that one time sensitive deal and that one time sensitive opportunity is on.

Something More about Existence

Existence exists because she cannot stop from existing. Existence has no choice but to only exist. Existence is composed of three portions of lack but which can be seen and existence also is composed of one portion which is but cannot be seen as of yet and which makes for three or even makes for four or for more portions…

Existence is composed of opposites which really are more opposites of themselves than to the opposites to which they are opposites.

And this opposition or contrast which makes possible for existence to attract herself and it appears that existence moves or is in motion because of the very attraction of herself.

And existence is known or is seen more through the little which is seen of her and through the more which is seen of her, less of her is known or less is seen…

And even though existence exists and she attracts herself and she also is eternal knowledge and also she gives knowledge to the vacuum of space or to the universe and thus creating the beginning or the times, existence does not know that she exists even though she also renews herself so that she can continue existing and existing as if always new and as if nothing has happened and will never happen…

But existence renews herself or is reborn when the vacuum of space becomes empty again or when the knowledge given to the universe is not taken as acknowledgement and she once again gives or once again throws more knowledge into the vacuum of space or into the space of the universe, thus making as if a new beginning and everything which was before as if it never had being or as if it never were…

But that ability which existence has of renewing herself and of continuing for all eternity without knowing end or ending thus we conscious beings also possess.

But we are not obligated to live eternity as neither we are obligated to death or to the end or the ending. Death or the end or the ending comes because it is not done to continue but to continue as if new and to continue forever or to continue for all the times…

Those that do not want eternity thus they just wait for death and death will do for them so that there is no eternity for them…

But those of us that want to continue with life without seeing or without knowing death thus we must do to be reborn and that rebirth or being born again truly is done with knowledge or with acknowledgement.

Thus, we must give knowledge or give acknowledgement to that grandiose part of existence which is but which cannot be seen as of yet but even so it is what gives or it is what grants knowledge not only to the vacuum of space or to the universe but also it is what gives or what grants knowledge to every conscious being that requests it…

And just as one presents oneself to that grandiose part of existence which gives or which grants or which even lends knowledge, thus that grandiose part of existence will be to one.

In other words, that grandiose or glorious part is the Creating part of existence or the renovating part or the part which gives identity to all existence or which reacts with all the opposites of existence…

Thus, he that gives or that grants or that lends to the Creating part or to the Renovating part of existence the knowledge or the acknowledgement of God or of Creator or of Renovator thus he also will have the knowledge or the acknowledgement of God or of Renovator and God will present to him or come to him or will allow him to draw near with that very same knowledge or acknowledgement…

Existence gives or grants or lends to the vacuum of space or to the universe knowledge in the form of matter or stars or light.

And if that knowledge in the vacuum of space or in the universe is converted into acknowledgement thus the universe will continue as if forever new without ever knowing end or ending nor knowing or remembering beginning…

But if that knowledge given or granted or lend to the vacuum of space or to the universe does not renew or is not converted or is not transformed into acknowledgement, thus that knowledge in the form of matter or stars or light thus will lose her energy and there will only remain and will be space dust no matter how large the piece of space dust and there will only be darkness or there will only be emptiness or vacuum…

But that does not remain like that because now the dust which remained in the vacuum of space or in the vacuum of the universe must be taken out to give or to grant or to lend new knowledge to the vacuum of space or to the universe and that new knowledge will make a new beginning, in where there will not be any memory that there ever was a beginning before…

The manner in which existence takes the dust from the vacuum of space or from the universe is with black holes which truly are black vacuum cleaners.

The black vacuums or black holes also suck up any other matter or star which has remained in the vacuum of space or in the space of the universe…

Once there no longer is matter or dust in space thus the black holes also will turn of or will stop from functioning because of lack of energy and they will disperse or they will disintegrate in the vacuum of space.

This emptiness in space, which now has become as if a vacuum or new space because of becoming as before the beginning thus will attract new knowledge in form of matter or stars or light…

And if in that new beginning the same happens which happened in the first, in where there was no renovation or there was no acknowledgement or no rebirth to continue, thus also will have its end or ending even though it may take billions of years…

But that does not have to be as the above because as long as there are conscious beings in the universe, the universe has the very grandiose opportunity of renovating or of rebirth or of receiving acknowledgement so that the universe because of the conscious beings the universe will continue without ever knowing end or ending…

Existence without knowing it renews every time she lends knowledge in the form of matter or in the form of stars or of light to the vacuum of space or to the universe and the universe cannot continue for lack of acknowledgement or for lack of matter or for lack of renovated energy and thus the universe comes to its end or to its ending and thus making space for another beginning which will become as if the first beginning and also as if it never had an end or never an ending before…

But if the conscious beings are reborn or revive or take new life or receive acknowledgement through the very same knowledge or acknowledgement which they give or grant or lend, the universe will never ever have end or ending because the universe will continue as if forever new and as if it never had any beginning…

Now then, every conscious being truly has the very grandiose opportunity of being reborn or of reviving or of taking new life or of receiving acknowledgement or rename to be able to continue eternally with life and in complete harmony and in complete abundance...

But if the conscious being does not desire that very grandiose opportunity of living eternally and living in complete harmony and in complete abundance thus that conscious being only has to wait to die and that will he his end or his ending and nothing will become of that conscious being because eternity or immortality is not obligated or is not imposed, even a rock will stop from being or from existing...

Thus, one needs to be alive and conscious to be able to have or can receive immortality in the form of salvation and with her continue alive renewing and renewing also everything else as savior...

Just as knowledge in the form of matter filled with energy or in the form of stars or in the form of light enters the vacuum of space or enters into the universe, thus that same way also thought or knowledge or illumination enters the conscious mind.

That thought or that knowledge or that illumination can take the conscious being to a great state or from one state to another state or to a greater identity or from one identity to another identity even though that conscious being really continues with his physical form but every time that that conscious being enters into a greater state of knowledge or into a greater or new identity because of his knowledge, thus the physical form of that conscious being also is refreshed or is seen as if a new form...

But if the thought or if the knowledge or the new identity which enters in that conscious mind of that conscious being is a limited thought or is a limited knowledge or is a limited identity, thus that thought or knowledge or that identity, even though some type of energy or be it negative or be it positive, does not take that conscious being very far or into a greater state of identity and that thought or knowledge or that identity will disperse and the conscious mind becomes once again as if empty.

And if that conscious being nothing does with his conscious mind to have thought or knowledge or identity so that the thought or the knowledge or the identity takes him to a greater state or to a greater identity in where not only his conscious mind will be refreshed or becomes as if a new mind but also his physical form or body also will refresh or even could be cured from certain lacks or faults, such as of that of deafness in one ear or both and also some emotional lack or fault such as loneliness or shyness…

But if the conscious being in the course of his life does not enter into a greater state of thought or of knowledge or of identity, thus the conscious being keeps on dying until he completely dies and his body will discompose until it turns to dust and the conscious being has lost his the very grandiose opportunity of rebirth and of continuing with life as if new in complete or in perfect harmony and also in complete or in perfect abundance, perfect because it will be an abundance which will never ever end…

Now then, once the universe stops from discomposing or no longer the vacuum of space ever returns to nothing because of the conscious being coming to their maximum state or coming to their maximum identity and that way keeping the universe practically alive, thus there no longer will enter more knowledge in the form of matter or in the form of stars or of light because now that makes it possible the conscious beings because they will be the matter or the stars or the light or the illumination of the universe because of they being illuminated until the maximum or until perfection…

n other words, there will no longer be any more beginnings nor there will be anymore ends or anymore endings and the universe will become as if there were never beginning because of the universe becoming as if new for all of eternity…

And all the different parts of existence will act or will react as if one as the same the body and the conscious mind of the conscious being will act or will react as if one or as a single part and existence will be one with the conscious being because of the conscious being becoming or as being existence herself and reflecting through his body her glory…

Thus, when one as seed for more came out from the entrails of a man and one entered into the entrails of a woman, one had no memory of that exit or entrance even though one as a seed was in harmony and in abundance in those entrails and one also came out with all gladness and with all joy and also with all feeling of abundance and entered into the entrails of a woman and there also sought for knowledge of life to life receive and in her also enter and once in her also one forgot because once again one entered in harmony while one was transformed or one took the form of life which one did for or for the one which one received through the act of one or because of one's movement to find life…

And when one became complete in those new entrails, one humbled and one took the grandiose position of contender and one came out or one entered into the entrails of the world not only as much more but also one came out or one entered for much more.

But in the world one did not remember that one came out from harmony while one keeps completing the form of contender which could take one to not only come to be conscious but also which could take one to rebirth and continue with life as if with a new form; but if there were no rebirth because of lack of knowledge or because of lack of identity, thus that form not reborn would take one to death and that would be the end or the ending of all of her, life, and also of all of one…

The More Understanding of Existence one has

The more understanding of existence or of reality one has of reality or of existence, therefore, the greater existence or reality will be to one as one will be to reality or to existence.

Reality or existence or the universe has to do with knowledge and acknowledgement, even unseen reality has to do with knowledge and acknowledgement even if negative knowledge or acknowledgement.

Space is an unseen reality or unseen knowledge or acknowledgement but space can be acknowledged but only when there is knowledge to acknowledge space with.

In other words, space can really be acknowledged through or because of matter or light. And matter as well as space can be divided into three major parts. Space is not only emptiness or is lack but space is also dark and cold. These three parts which make space are lack or are negative (-) knowledge or negative (-) acknowledgement. Positive knowledge or acknowledgement, therefore, is matter and its three major parts, such as weight or mass, light and heat. And matter only interacts with space while in the vacuum or emptiness of space.

Now then, existence is about knowledge and acknowledgement or confirmation and reconfirmation or on and off or 01.

When matter which is knowledge enters the vacuum of space, space begins to react with matter as a form of acknowledgement and the vacuum of space begins to compress matter until matter explodes and it begins to expand in the vacuum of space. But the friction which matters has from space makes matter to last less in the vacuum of space.

Now, before matter entered the vacuum of space, space was in a state of tranquility or space was neutral or 0.

But when matter entered the vacuum of space, space acted as if negative due to the interaction with matter such as the compression of matter but when matter exploded and began to expand through the vacuum of space, space began to act as if positive.

But when matter begins to turn off due to lack of energy, space also begins to act negative.

And when black holes begin to appear due to the collapse of large stars, the vacuum of space begins to act even more negative.

And when there no longer is matter in the vacuum of space because the black holes vacuumed all the matter up, the vacuum of space has become completely negative due to the black holes now expanding throughout the vacuum of space in search of matter.

But just as matter ran out of energy, so will the black holes run out of matter and collapse or disperse and the vacuum of space will once again become neutral or return to 0 and once again become ready to receive matter or new knowledge or 1.

But this new beginning is as if the very first beginning because there will not be any memory that there ever was any other beginning or beginnings.

Also, the vacuum of space is dimensional and will receive matter or knowledge or 1 up 118 times.

That is to say, there is 118 beginnings which really begun at the same time or spontaneously and enter into 118 dimensions or space time.

But not only that, matter or knowledge or 1 enters the first dimension or space time as 118 percent and that 118 percent equals or adds back to one.

In the same manner, 2 equals 236 percent and that 236 percent equals or adds back to one and so forth with the other numbers up to 118. That is, 3 equals 354 percent and that 354 percent equals or adds back to one.

And finally, 118 or the last number equals 13924 percent, which is 118 times 118, and that 13924 percent equals or adds back to one.

The dimensions of time also add up as if they were one. And they increase in size from one to 118 times.

That is, the first is one but the second is twice as the first and the third is three times as the first but they still all add up as if they were just one dimension.

And if the speed of light was ever measured correctly the speed of light would be equal to one. The speed of light is about 186, 282 miles per second which adds up to 9 or point nine or 09. This loss of 1 or of 01 is due to the compression of the vacuum of space. It is similar to water when it freezes. When water freezes water loses energy and weights from 1 to point 9.

Thus, existence or reality is one or existence or reality is knowledge or more like 01 which really means knowledge or acknowledgement.

For the conscious being, therefore, this information is very important because for this information the conscious being unknowingly struggles or contends for to be able to enter into higher or taller or greater conscious modes or be truly illuminated and through that illumination enter into a higher or taller or greater conscious existence and with this greater conscious existence or identity the conscious being can mode or transform existence or reality or his natural environment according to his will.

Of course, that forming or transformation really is done through his mouth and through the use of the spoken word or the zero plus one factor or language, more like binary because of the presentation or the knowledge and the acknowledgement.

The Plus Zero Negative Factor

Those that accept an idea or ideology blindly or without trying or testing that idea or ideology, thus they unknowingly become liars.

But nonetheless, liars they are and as liars they live and will lie to others to convince them to accept or believe and they will keep on lying until death herself and death herself will close their lying trap.

Now, when one does the movement or gets interested for knowledge, thus one truly becomes that knowledge. When the universe or creation came to exist, the universe or creation came to exist because of knowledge.

The proof is in the numbers or in matter which is composed of elements and the elements in turn are really numbers, real numbers.

In other words, matter is knowledge because matter is composed of elements and the elements are composed of numbers and the numbers are composed of positive, neutral and negative states, thus the, (+ 0 -), plus zero negative factor or a magnet with its three opposites parts.

What this really means is that knowledge can become neutral or void or useless and then become negative or contra productive if at first nothing was done with that knowledge, such as using or converting knowledge into acknowledgement or a positive or useful respond.

In the same manner as above, the universe or the vacuum of space becomes positive when matter enters the universe or the vacuum of space.

But as matter begins to burn out, the universe or the vacuum of space begins to turn neutral until it turns negative, negative because of the black holes which now rule the universe or the vacuum of space which they now also suck up any remaining dust or matter to make space for another beginning.

But this new beginning is as if the very first beginning because there will not be any trace that there was ever a first beginning.

But the above matter would only be a theory or an idea or ideology if the conscious being, which is also real knowledge, did for real acknowledgement and the conscious being would have the power and the authority over matter to refresh matter and thus keep the universe always positive and refreshing in double abundance, all five portions of her!

Conclusion

Can you, the interested and the now glad and joyful reader, gladly and joyfully truly imagine what would had happened if the simple and the truthful or the real question as to the physical origin of the universe or as to the physical origin of all Existence, or even the truthful or the real question as to physical creation, were simply asked and simply answered when conscious thought first entered the human conscious mind?

Really, the physical world or physical creation would not be the physical and the mental lack and the physical and the mental chaos that the world or that physical creation really is!

There would had never been insignificant or useless ideas or useless points of view that, unfortunately, still today bring forth false, half-truths or useless knowledge, not allowing mankind to further expand his conscious mind into simple enlightenment or simply into completeness or perfection, and therefore, not allowing the future or complete or perfect Existence or perfect creation here and now.

Unlike the divisible atom, where the electrons cannot naturally return to the center or to the nuclei of the divisible atom or where the proton naturally cannot occupy electron rings or the electron field or shield, and certainly the physical universes would really be physically different and therefore all of actual Existence, the conscious mind or the conscious living being can deter, can reverse, can change or really can improve the conscious mind's or the conscious living being's point of view, believe or understanding or the way of conscious thinking or the way of conscious functioning or even the way the conscious living being learns simply by changing or improving conscious thought or even by improving mood or by simply acknowledging or by simply seeking to be enlightened or to be reborn!

By getting really interested or by improving one's mood is a very simple way to improve one's conscious thought or one's point of view. One can see more by simply getting interested or simply by being glad or joyful or even excited. Getting interested or getting glad or joyful or even getting excited is a major part of learning, but getting interested or getting glad or joyful or even getting excited is not an automatic process.

The conscious living being also has to make an effort to get interested or to get glad and joyful or even excited.

In other words, the conscious living being really has to make an unnatural conscious effort to learn how to really learn. Once the conscious living being simply or actually knows or simply or really understands how to learn or really understands the real reason for really learning, which is really recollecting or simply remembering, the conscious living being will know or will understand how.

That is to simply say, once the conscious living being really knows or really understands why or really understands what or really understands the simple reason for getting interested or the simple reason for getting glad or even excited, the unnatural process of getting interested or the unnatural process of getting glad or even getting exited then becomes automatic, and the conscious living being through conscious thought or conscious effort will really or will actually know how (to complete Existence or how to complete actual reality or physical creation).

Interestingly, joy, gladness or happiness, as really is also conscious thought, is pure energy. A joyful, a glad or a happy thought can really or can actually keep the conscious living being a good number of days without sleep, without hunger and without feeling tired or stressed out. Only when the joyful, the happy or the interested thought is gone or completed or has actually become natural, the conscious living being once again begins to need sleep, to need food and to need rest, both physically and mentally.

True, complete or real joy; true, complete or real happiness; or true, complete or real interest really brings to light physical things or the reality unseen. True, complete or real joy or real happiness is not an automatic or is not a natural process at first, even though true, complete or real joy already exists. And because mankind instinctually or naturally knows that true joy or that true happiness is or already exists, mankind seeks true joy or true or real happiness.

However, that is temporary or that is natural joy or that is temporary or natural happiness. That is the reason mankind really feels miserable or really gets into trouble looking for temporary or natural joy or temporary or natural happiness.

True, complete or real joy or happiness is unnatural, and therefore, really requires conscious effort or conscious thought. However, this really has nothing to do with positive thinking. Positive thinking is only a very simple point of view. Positive thinking is not really or actually a good thing.

Positive thinking can really be a real hell or a very heavy burden. Mankind usually, as a creature of unnatural habit, imitates mostly what mankind really likes or what really brings mankind the most pleasure or the most joy.

Changing or improving conscious thought, changing or improving behavior or changing one's way of thinking or changing point of view is unnatural and is also not very easy. It really requires a lot of effort, a lot of energy and a lot of time for a lot of nothing.

Positive thought really is, therefore, one of those good sounding lies or good sounding half-truths or false science. Positive thinking bears all as it pretends to dress and nurture.

But once the conscious living being or mankind through conscious effort really acknowledges, realizes or really understands or really recollects the real or the actual reason for true, complete or real joy or real and lasting happiness, then true, complete or real joy or real and lasting happiness simply becomes an unnatural but automatic process as conscious thought once became an unnatural but automatic process as the very simple magnetic field or the very simple neutral shield always was an automatic process of physical existence or of physical reality.

In short, once the unnatural but automatic process of true, complete or real joy or real and lasting happiness kicks in, new energy is felt both in body and in mind. At first that new energy will feel like an ecstasy or a natural super high.

In fact, it is really or actually the first enlightenment! It is a true, a complete or a real joy or a real and lasting happiness that keeps coming, even without further effort from the conscious living being, leading to a second and final enlightenment or final acknowledgement. The very simple anticipation of the next enlightenment really or actually is what makes possible the new found joy or real and lasting happiness.

Ironically, when true, complete or real joy or real and lasting happiness begins automatically as once did conscious thought, the conscious living being really does not understand where it really comes from and so the conscious living being begins to fight real and lasting happiness, as the conscious living being once fought against conscious thought, by feeling odd or by feeling strange, and thus begins to reject that new and automatic and real joy or real and lasting happiness.

As it has been seen and as it has been proven, the universe, Time or simply all of Existence or even physical creation is a product of two simple but infinite magnets in the very simple form of two infinite cubes: The Space Cube and the Matter Cube or call the two infinite magnetic cubes Reality and Non-reality.

By using the Periodic Table of the Elements, which is really a very simple diagram of a bar magnet or a simple diagram of an electric motor or a generator, the above statement, that simply states that all of Existence or that physical creation is really a simple product of two simple but infinite magnets, has simply and thus been proven.

And also, by using the simple and the divisible atom, which really is also a very simple microscopic diagram of a bar magnet or an electric motor or a generator, the first statement of this simple conclusion or this simple continuation has thus simply been proven.

In other words, the true question as to the physical origin of the expanding universe or physical creation has really been simply answered!

And in conclusion or in continuation, the most important thing in all of creation or in all of Existence, seeing or not seeing, once again, simply is or actually is but conscious thought!

Conscious thought only exist in conscious living beings. Conscious thought really is not a natural or is not an automatic process as is all of Existence or as is reality.

Even all of Existence or all of creation as well as non-existence really require something. Conscious thought is not a natural process and conscious thought really requires conscious effort from the conscious living being. Even though knowledge simply exists or simply "the fact simply and already but is," the conscious living being does not know or does not learn or does not really recollect or does not really remember and thus simply cannot complete Existence or physical creation until the conscious living being makes a conscious effort to simply think and to simply ask that unnatural but true or real question which will simply bring the conscious living being complete or perfect knowledge or acknowledgement or enlightenment!

And the real or the actual purpose of complete or perfect knowledge is simply to let the conscious living being know or really acknowledge that the conscious living being but is and that the conscious living being can really do more with what the conscious living being could really be or could really recognize or could really recollect or could really remember!

Existence as knowledge simply is or simply exists; but ironically, Existence or knowledge knows not. However, just because all of Existence or all of knowledge really knows but not, really does not make Existence or knowledge dumb or useless. All of Existence or all of knowledge acts or reacts through a simple magnetic or through an electronic impulse or acts or reacts through an actual magnetic instinct.

But knowledge does not know that knowledge really is or that knowledge simply exists. For example, the simple apple tree grows knowing not; but the tree's natural instincts, through knowledge already gotten or already present, is to produce a seed or to simply multiply itself.

The apple of that apple tree ripens or temporarily becomes almost complete not knowing that it does, so that the seed in the almost complete apple can be carried away, but not by the wind or not really by rolling on the ground even though the apple is almost round, but really carried away by living beings. That is the very simple or the actual reason for the sweetness or the sugar in the almost round apple!

The simple apple in the belly of the living beast or in the hand of the living being is simply or is really a continuation, and not the conclusion, of something that already was. And even though the apple tree which is really but simple knowledge was really before the seed, the purpose of the apple tree was always the seed, an extension or an expansion of physical knowledge that really already was or was already present!

Another very interesting thing about the knowledgeable apple seed is that the apple seed not only carries all of the physical information of the apple tree, but also the apple seed carries all the knowledge or all the information about the environment or the physical surroundings where the apple tree was or is. The knowledgeable apple seed no longer has to experiment with the environment or with natural surroundings since the knowledgeable seed already has that physical information. Instead, the seed as a new or as a second tree will gather more or new knowledge for the next apple or the third tree not yet present.

Whether the new or the second apple from the new or the second tree is sweeter or is more sour than the very first or the previous apple, simply depends, of course, on the physical reaction or the interaction that the living beast or the living being gave to the very first or to the previous apple.

The knowledgeable apple is often colored red to attract the eye of the living beast or the living being with hands, and the redder or the more reflective or shiner the apple, the better for the apple tree and thus the better for the knowledgeable apple seed.

If the natural or the physical surroundings are all red, then the knowledgeable apple with time simply changes to another but more attractive or more attracting color.

By the way, animal seeds or animal cells as do plants seeds have the very same ability or instincts to carry knowledge of previous generations. But because of the animal's irrational or emotional behavior, or even point of view or new physical location, that first or that previous knowledge or instincts is rarely re-used and usually is forgotten, rejected or even lost.

A very good or an excellent example of lost knowledge or lost instinct is the domestic animal. Left alone in an unfamiliar environment the domestic animal will not survive. Even house plants lose the ability to survive because of the lack of information or the lack of knowledge of the new environment or new surroundings.

Once again, conscious living beings are really the most important thing in all of Existence, and that is what Existence really reflects or that is what all of Existence really wants or really requires the conscious living being to know. But not only that!

All of Existence also requires, also wants, or also needs, as the apple tree, the conscious living being to simply act or to react to knowledge, so that all of Existence as well as the living conscious being can be further completed or perfected through conscious thought.

In a way, conscious living beings are actually or are really inside a very large mechanical or infinite magnetic brain or inside a very large electronic brain; and conscious living beings are really the thinking brain cells or living knowledgeable seeds or even living knowledge processors.

Ironically, Existence will not be complete or will not be perfect until all the needs of the conscious living being are simply but met through conscious thought. And that is the actual reality!

The only true or the only complete or the only perfect or real beginning in Time and in all of Existence or even creation is the simple or the actual beginning of conscious thought. The only limits that the human race, the conscious living being or that mankind really has are the very same limits that the human race, the conscious living being or that mankind really gives or really sets upon himself.

So, therefore, let's get glad and joyfully continue to think and rethink!

Thinking or conscious thought will one day keep the universes or creation as well as mankind from collapsing. Conscious thought will be the conscious shield that will keep the simple universes or creation from collapsing or from imploding into a giant black hole or into a giant grave.

Conscious thought will be the conscious shield or the conscious energy that will one day really keep the simple human conscious mind or the human brain from physically collapsing or from physically imploding also into a hole or into a grave, thus saving the physical universes or saving creation and thus completing or really perfecting physical Existence or physical creation!

The final reality of the matter is that as long as there is conscious thought in the physical universe or physical universes, the physical universes or physical creation will not collapse or will not implode or will not even turn on itself.

The reality of the matter is that there was only one physical creation or only one physical beginning, actually a physical continuation of what already was.

As long as there is conscious thought, only one physical beginning or only one physical continuation or only one physical creation will be so until infinity!

Something about the Author

I truly hope that you enjoyed reading this very short work as much as I truly enjoyed writing it.

I was born in the Caribbean Island of Puerto Rico, where I am now living with my youngest of two sons.

As a child and most of my adulthood, I lived in the States. I mostly lived in the City of New York, mostly in the South Bronx, sometimes called Fort Apache.

I went to Marist College in the town of Poughkeepsie, Up State New York, where I studied some Chemistry; and later transferred to Hunter College, in the City of New York.

But after quitting school I worked in two different chemical plants in the state of New Jersey. In 2008, I moved here back to Puerto Rico.

I am one of those, and perhaps the only one in these very difficult times, who with complete sincerity seeks the truth the truth of the universe, the truth of the creation of God and the truth of what is the salvation of man and man, as far as I know, will not be saved from eternal death by the technology of man as neither the very sword will save man.

Technology as the sword will only save man but save man only temporarily. And technology can also be a double edge sword or contra productive because man does not really use it to expand or to increase his time or to expand or to increase his conscious mind or to truly increase his real knowledge, which will give him a greater identity.

My search for the truth of the universe, for the truth of the creation of God and for the truth of what is really the salvation of man, has done to me very good, so good in fact that it has taken me to not only achieve or to have real faith of God, but also my search has allowed me to really understand what is in truth the salvation of man by God the good and loving Creator.

And the very simple truth is that without real knowledge of God, thus there will not be the salvation of God and without the salvation of God, thus there will neither be the salvation of man nor will there be the salvation of the universe, because as man will return eternally dead to the very dust, thus also the universe will return to nothing…

Thus, in my very simple writings there are certain truths or real knowledge of life!

Very glad and very joyful he who in truth lends attention!

Who am I really?

My pen name as a writer is Francisco Sanchez, Jr. and I am truly on a very grandiose crusade of rebirth alive or to be born again into royal life with complete gladness and with complete joy and also with complete abundance of God but as much more than God and as much more than Creator.

Now then, one who truly is on a very grandiose crusade cannot follow another or cannot let himself be surrounded by his beloved ones or his fans because he cannot cross over them or he cannot cross over because of them being in the way or because of them blocking the path which is but which cannot be seen until rebirth or until one is born again.

I do not ask to be followed, not because I will not lead, but because I will not look back but I will look to my right and to my left to see who walks with me.

But those that truly decide to follow me will become as me and as me will truly receive or gather true knowledge because my struggle or contention or my very grandiose crusade of rebirth is true, so true in fact that I have become a much better person because of the true faith which I have come to receive through my search and research for the truth.

And because I have come to have true faith or faith of God, thus I use my true faith as a shield to repel or to reject other beliefs or good sounding lies!

Therefore, to rebirth alive or to be born again while still living here on the very earth which will be as in the very heavens through rebirth!

$$(((((((0+1)))))))$$

If you truly enjoyed this simple and humble work, please leave a comment according to your good pleasure and give also a rating but also according to your good pleasure.

Thanks so very much for your time and best of wishes, Francisco Sanchez, Jr.

Thanks for reading my work!

0+1 = peace and knowledge

And may there be peace and knowledge to every man that truly seeks her!

www.ingramcontent.com/pod-product-compliance
Lightning Source LLC
Chambersburg PA
CBHW030654220526
45463CB00005B/1770